W9-BAX-397

"Edward Slingerland is one of the world's leading comparative philosophers and the foremost advocate of bridging the gulf between cognitive science and the humanities. In *Trying Not to Try*, he reminds us that philosophy truly is a way of life, that classical Chinese philosophy offers deep insights into human flourishing, and that this classical Chinese wisdom anticipates in compelling ways what the best contemporary cognitive science teaches. This is a landmark book—clear, sparkling, and humane."

—Owen Flanagan, James B. Duke Professor of Philosophy at Duke University and author of *The Bodhisattva's Brain*

"A fascinating read. With state-of-the art science and interesting stories, Slingerland provides key insights from the East and West for achieving happiness and well-being."

—Sian Beilock, Professor of Psychology at the University of Chicago and author of *Choke*

"Through a combination of hard science and ancient philosophy, *Trying Not to Try* has convinced me that my usual approach to life—smashing through walls and grinding out painful victories—isn't all it's cracked up to be. Sometimes trying hard is overrated. Slingerland has written a charming, intellectually rigorous book that can help all of us improve our lives."

—Jonathan Gottschall, author of *The Storytelling Animal*

"Slingerland's book exemplifies the very principles it elucidates. Although the material is sophisticated, we effortlessly glide through a highly original integration of ancient wisdom and modern science toward a deep understanding of how one can simultaneously set a course in life and live spontaneously."

—Jonathan Schooler, Professor of Psychological and Brain Sciences at the University of California, Santa Barbara

"This wonderful book not only shows us how to live a more satisfying life, it helps explain why social life is even possible: spontaneity, Slingerland argues, is the key to trust and, ultimately, the evolution of cooperation. A thought-provoking book by a truly gifted writer."

—Harvey Whitehouse, Director of the Institute of Cognitive and Evolutionary Anthropology at the University of Oxford

"I tried hard to avoid reading this book—just too much to do. But I lost control, dipped in, and was swept along by apparently effortless prose describing the contrast between Confucianism and Taoism, and its relevance to our modern lives, including the good evolutionary reasons why commitment is usually more successful than manipulation. This is the perfect book club book."

—Randolph Nesse, Arizona State University Center for Evolution, Medicine, and Public Health; author of *Why We Get Sick*

"A remarkable time-traveling synthesis that shows how classic Chinese philosophers anticipated contemporary brain science and also looked beyond it, offering sage advice about how to live lives that flow. We meet Confucius, Daoists, the first Zen master, a sixth-century hippie, and other ancient Eastern educators, whose ideas have never been rendered more relevant to our times."

—Jesse Prinz, Distinguished Professor of Philosophy and Director of the Committee for Interdisciplinary Science Studies at the City University of New York

TRYING NOT TO TRY

THE ART
AND SCIENCE
OF SPONTANEITY

Edward Slingerland

CROWN PUBLISHERS
New York

Copyright © 2014 by Edward Slingerland

All rights reserved.

Published in the United States by Crown Publishers, an imprint of the Crown Publishing Group, a division of Random House LLC, a Penguin Random House Company, New York.
www.crownpublishing.com

CROWN and the Crown colophon are registered trademarks of Random House LLC.

Grateful acknowledgment is made to *Sports Illustrated* for permission to reprint "How It Feels . . . To Be on Fire" edited by Kostya Kennedy, February 21, 2005. Copyright © 2005 Time Inc. All rights reserved. Reprinted by permission of *Sports Illustrated*.

Library of Congress Cataloging-in-Publication Data
Slingerland, Edward G. (Edward Gilman)
   Trying not to try : the art and science of spontaneity / Edward Slingerland. — First Edition.
       pages cm
   Includes bibliographical references and index.
1. Philosophy, Chinese. 2. Spontaneity (Philosophy). 3. Nothing (Philosophy). 4. Struggle. I. Title.
   B126.S6453 2014
   181'.11—dc23

                                        2013023431

ISBN 978-0-7704-3761-9
eBook ISBN 978-0-7704-3762-6

Printed in the United States of America

Book design by Nicola Ferguson
Jacket design by Oliver Munday

10 9 8 7 6 5 4 3 2 1

First Edition

Contents

TRYING NOT TO TRY

Introduction

THERE IS A WONDERFUL GAME AT MY LOCAL SCIENCE MUSEUM called Mindball. Two players sit at opposite ends of a long table. Each wears a headband equipped with electrodes, designed to pick up general patterns of electrical activity on the surface of the brain. Between the players is a metal ball. The goal is to mentally push this ball all the way to the other end of the table, and the player who does so first wins. The motive force—measured by each player's electrodes, and conveyed to the ball by a magnet hidden underneath the table—is the combination of alpha and theta waves produced by the brain when it's relaxed: the more alpha and theta waves you produce, the more force you mentally exert on the ball. Essentially, Mindball is a contest of who can be the most calm. It's fun to watch. The players visibly struggle to relax, closing their eyes, breathing deeply, adopting vaguely yogic postures. The panic they begin to feel as the ball approaches their end of the table is usually balanced out by the overeagerness of their opponent, both players alternately losing their cool as the big metal ball rolls back and forth. You couldn't wish for a better, more condensed illustration of how difficult it is to try not to try.

In our culture, the benefit of not trying too hard—of "going with the flow" or "being in the zone"—has long been appreciated by artists. The jazz great Charlie Parker is said to have advised aspiring musicians, "Don't play the saxophone. Let it play you." This same openness is also crucial in acting and other performing arts, which fundamentally rely on spontaneity and seemingly effortless responsiveness. A stand-up comedian who is not in the zone is not funny, and an actor who is not fully inhabiting his or her role comes across as wooden and fake. Explaining how to prepare for a role, the actor Michael Caine cautions that simply memorizing the script and trying to act it out step by step will never work; when it comes time for your line, the only way to bring it off authentically is to not try to remember it. "You must be able to stand there *not* thinking of that line. You take it off the other actor's face. He is presumably new-minting the dialogue as if he himself just thought of it by listening and watching, as if it were all new to him, too. Otherwise, for your next line, you're not listening and not free to respond naturally, to act spontaneously."

The importance of being in the zone is perhaps nowhere more appreciated than in professional sports, where the competitive edge provided by complete absorption is the stuff of myth. A 2005 piece in *Sports Illustrated* consists solely of quotations from professional basketball players about what it feels like to be on fire:

> There are books you can read about how to get into that shooting zone, how to prepare yourself, but it's never something you can predict. The ball feels so light, and your shots are effortless. You don't even have to aim. You let it go, and you know the ball is going in. It's wonderful . . . It's like a good dream, and you don't want to wake up.
>
> —Pat Garrity, Orlando Magic forward

> It's like an out-of-body experience, like you're watching yourself. You almost feel like you don't even see the defense. Every move you make, you feel, God, that guy is slow. You're going by people. You don't even hear the regular noise you hear. It's muffled. You go to practice the next day, and you say, "God, why can't I do that every night?" Guys have wanted to bottle that feeling.
>
> —Joe Dumars, former NBA All-Star guard

The reason professional athletes would love to bottle that feeling is that it all too easily disappears. As Garrity says, professional basketball players in the shooting zone don't want to wake up, but they often do. Ben Gordon, formerly a guard for the Chicago Bulls, puts it like this: "When the feeling starts going away, it's terrible. I talk to myself and say, 'C'mon, you gotta be more aggressive.' That's when you know it's gone. It's not instinctive anymore."

Falling out of the zone is terrifying, and athletes try to avoid it at all costs. The history of sports is full of stories of otherwise promising athletes who somehow lose their mojo and then disappear into obscurity—or, perhaps worse, become famous precisely *for* having lost their stuff. Baseball fans are familiar with "Steve Blass Disease," named after a superstar pitcher for the Pittsburgh Pirates in the 1960s and 1970s. After almost a decade of effortlessly intimidating the best players in the world, Blass suddenly began to lose the ability to perform in actual games. He continued to be fine in practice. He'd suffered no injuries, lost none of his actual physical skill. He just could no longer *throw* when it really mattered. Armies of sports psychologists analyzed him, coaches tried beating it out of him through grueling exercise routines, but nothing worked, and Blass was eventually forced into early retirement.

An inability to relax into the zone is also a danger for artistic performers. One famous example is the pop singer Carly Simon. She had always been reluctant to be up in front of a crowd, but her stage fright came to a head at a concert in 1981, when the tension caused her to freeze. "After two songs, I was still having palpitations," she later told a reporter. "I suggested that I might feel better if someone came on the stage. About 50 people came up, and it was like an encounter group. They rubbed my arms and legs and said, 'We love you,' and I was able to finish the first show. But I collapsed before the second show with 10,000 people waiting." Simon's onstage meltdown led to a long hiatus from the public eye, although—unlike Blass—she later managed to make a professional comeback.

There is a widespread recognition that this tension—how to force yourself to relax, to shut off your mind when you need to—is a challenge for professional athletes or other performers. For those functioning at an elite, competitive level, spontaneity is a basic job requirement; their livelihood depends on the ability to reliably get into the zone. What fewer people realize is that this is a challenge we *all* face. We may not be subject to the same public pressures as Steve Blass or Carly Simon, but in many ways our lives can be seen as a massive game of Mindball.

The pervasiveness of the problem is perhaps more obvious when it comes to physical activities. Even casual athletes or performers are familiar with the pain of falling out of the zone or finding it just beyond their grasp. Imagine that you are in the final set of a tennis match—playing your best game ever and about to defeat your former intercollegiate tennis star friend for the first time in recorded history—and the dawning realization that you are about to win makes you begin to lose. You become tense, overly cautious. You begin to *think* about your swing instead of just swinging, and your friend begins to close the gap. You know what you need to do: just *relax* and get back into the groove. The more you think about relaxing, though, the more you tighten up, and you watch helplessly as your lead disappears and your friend gets to gloat once again.

Or imagine that you are taking an introductory salsa class, and your initial awkwardness is exacerbated by an annoying instructor who keeps telling you to *be spontaneous*. "Relax! Have fun with it!" she chirps, as you stumble through the steps you've been taught and do your best not to crush the feet of your partner. "Relax! *Feel the music!*" The more she urges you to *have fun* the more tense you become. You find that a generous quantity of tequila helps with the relaxing side of things, but only at the cost of radically diminished motor coordination. Simultaneously skillful *and* enjoyable salsa dancing seems forever out of your reach.

Getting the mind to shut off and allow the body to do its thing is clearly a challenge. An even bigger problem—and one we encounter much more often—is the trick of getting your mind to let go of *itself*. This is the central problem in Mindball, where you can win the game only by relaxing, which seems to mean that you can win only if you don't *try* to win. In our everyday lives, this tension is perhaps most intensely distilled in the throes of insomnia. You have a big meeting tomorrow and need to be at the top of your game, so you go to bed early and try to relax into sleep, only to find yourself tortured by incessant thoughts, helpless in the grip of the restless monkey-brain. Counting sheep just makes it worse, no position seems comfortable; you feel in your bones how tired you really are, but how do you make your brain shut off? *RELAX!* you think to yourself, but it's no use.

Insomnia is a fairly simple case, but the problem also manifests itself in more complex—typically social—situations, where the impact is far greater. Consider dating. Anyone who has ever been single for a significant amount of time is familiar with the "never rains but it pours" phenomenon: you can sometimes go for long periods of being miserably alone, desperately trying to meet someone but having absolutely no luck. Then something happens, an encounter occurs, you go out, you have a great time, and suddenly it's raining women or men (or both, if you're so inclined). Attractive potential partners smile at you on the street, strike up conversations with you in cafés. The previously inaccessible beauty at the video store counter—who in the midst of your dry stretch would never even make eye contact with you—suddenly shows an interest in your predilection for Wim Wenders films, and the next thing you know you have plans for that Friday (a Friday!) to watch *Wings of Desire* and eat takeout Indian food. (This example is in no way autobiographical.) You sniff your clothes trying to detect any special pheromones you might be emitting, but if the phenomenon is biochemically based your senses are too dull to detect it. Bathing appears to have no negative effect.

Everyone enjoys these periods of deluge, but once you're back in a dry spell the pattern seems wasteful and fundamentally unjust.

There are too many potential dates when you can't enjoy them all, and none when they are really needed. Serious reflection—and during a drought you have a lot of time to reflect—suggests a possible reason for the pattern, or at least why it is so hard to consciously alter: the best way to get a date seems to be to *not want to* get a date. The problem is that it's hard to know what to do with this knowledge. How do you make yourself *not* want something that you actually *do* want?

For the most part, we—and by "we" I mean pretty much anyone with access to this book, inhabitants of modern, industrialized societies around the world—are preoccupied with effort, the importance of *working, striving,* and *trying.* Three-year-olds attend drill sessions to get an edge on admission to the best preschool and then grow into hypercompetitive high school students popping Ritalin to enhance their test results and keep up with a brutal schedule of after-school activities. Both our personal and our professional lives increasingly revolve around a relentless quest for greater efficiency and higher productivity, crowding out leisure time, vacation, and simple unstructured pleasures. The result is that people of all ages spend their days stumbling around tethered umbilically to their smartphones, immersed in an endless stream of competitive games, e-mails, texts, tweets, dings, pings, and pokes, getting up too early, staying up too late, in the end somehow falling into a fitful sleep illuminated by the bright glow of tiny LCD screens.

Our excessive focus in the modern world on the power of conscious thought and the benefits of willpower and self-control causes us to overlook the pervasive importance of what might be called "body thinking": tacit, fast, and semiautomatic behavior that flows from the unconscious with little or no conscious interference. The result is that we too often devote ourselves to pushing *harder* or moving *faster* in areas of our life where effort and striving are, in fact, profoundly counterproductive. This is because the problem of choking or freezing up extends far beyond sports or artistic performance. A politician who is not, at some level, truly relaxed and sincere while giving a speech will come off as stiff and uncharismatic—a problem that plagued U.S. presidential candidate Mitt Romney. In the same way, a real love of reading, a genuine commitment to learning, and a deep curiosity about the world cannot be forced. Like that most elusive of modern goals, happiness, spontaneity seems to be as tricky to capture and keep as the hot hand in basketball. Consciously try to grab it and it's gone.

WU-WEI ("OOO-WAY") AND DE ("DUH")

The goal of this book is to explore the many facets of spontaneity, as well as the conundrum it presents: why it's so crucial to our well-being and yet so elusive. In fact, the problem of how to try not to try is an ancient one, and it has engaged thinkers throughout history and across the world. Some of the most important and influential of them lived in early China. It is my belief that these thinkers, hailing from the so-called Confucian and Daoist schools, had deep insights into the human condition that can still prove very useful to us today. Looking at our lives through this early Chinese lens will require learning about two tightly linked concepts: the first is *wu-wei* 無為 (pronounced *oooo-way*), and the second is *de* 德 (pronounced *duh*, as in "no duh").

Wu-wei literally translates as "no trying" or "no doing," but it's not at all about dull inaction. In fact, it refers to the dynamic, effortless, and unselfconscious state of mind of a person who is optimally active and effective. People in *wu-wei* feel as if they are doing nothing, while at the same time they might be creating a brilliant work of art, smoothly negotiating a complex social situation, or even bringing the entire world into harmonious order. For a person in *wu-wei*, proper and effective conduct follows as automatically as the body gives in to the seductive rhythm of a song. This state of harmony is both complex and holistic, involving as it does the integration of the body, the emotions, and the mind. If we have to translate it, *wu-wei* is probably best rendered as something like "effortless action" or "spontaneous action." Being in *wu-wei* is relaxing and enjoyable, but in a deeply rewarding way that distinguishes it from cruder or more mundane pleasures. In many respects, it resembles the psychologist Mihaly Csikszentmihalyi's well-known concept of "flow," or the idea of being in the zone, but with important—and revealing—differences that we will explore.

People who are in *wu-wei* have *de*, typically translated as "virtue," "power," or "charismatic power." *De* is radiance that others can detect, and it serves as an outward signal that one is in *wu-wei*. *De* comes in handy in a variety of ways. For rulers and others involved in political life, *de* has a powerful, seemingly magical effect on those around them, allowing them to spread political order in an instantaneous fashion. They don't have to issue threats or offer rewards, because people simply want to obey them. On a smaller scale, *de* allows a person to engage in one-on-one interactions in a perfectly efficacious way. If you have *de*, people like you, trust you, and are relaxed around you. Even wild animals leave you alone. The payoff provided by *de* is one of the reasons that *wu-wei* is so desirable, and why early Chinese thinkers spent so much time figuring out how to get it.

The fact that no other language has a good equivalent to either *wu-wei* or *de* is quite illuminating and reflects a corresponding gap in our conceptual world. Just as the German loan of *Schadenfreude* has allowed English speakers to focus attention on an ever present but otherwise overlooked aspect of their emotional lives, I think that adding the words *wu-wei* and *de* to our verbal repertoires will help us gain insight into aspects of our mental and social worlds that we have tended to miss. Since I was introduced to these concepts as an undergraduate they've become a basic part of my vocabulary and have also spread quickly among my family members, friends, and acquaintances. "You're not being very *wu-wei* about this," my wife now chides me when I'm trying to force something that shouldn't be forced—a recalcitrant door, an obstructionist bureaucrat. "That guy just doesn't have *de*, but you do," I tell a colleague, in an attempt to explain why I want *her* rather than someone else to join me for an important potential grant interview. And she knows precisely what I mean.

Understanding these two concepts is essential to understanding early Chinese philosophy, which we'll explore by looking at five thinkers who lived and taught during the so-called "Warring States" period (fifth to third century B.C.E.) of China. This was a time of great social chaos and political upheaval. Powerful states swallowed up weaker ones, with the rulers of the losing side often put to the sword. Huge conscript armies roamed the land, devastating crops and making life miserable for the common people. It was also (not incidentally) a period of incredible philosophical creativity that witnessed the founding of all the major indigenous schools of Chinese thought. Despite considerable differences among these thinkers on a wide variety of issues—nature versus nurture, learning versus instinct—they all built their religious systems around the virtues of naturalness and spontaneity and felt that overall success in life was linked to the charisma that one radiates when completely at ease, or the effectiveness that one displays when fully absorbed. In other words, they all wanted to reach a state of *wu-wei* and get *de*.

They also all faced their own version of the Mindball challenge. How could they ask their followers to strive for a state of unselfconscious, effortless spontaneity? How does one *try* not to try? Doesn't the very act of trying contaminate the result? This is what I will be calling the paradox of *wu-wei*. All of our thinkers believed that they had a surefire way to resolve the paradox and get people safely into a state of *wu-wei*, as well as explanations for why their rivals *couldn't* do so. This was viewed as a particularly urgent problem because, for them, *wu-wei* and *de* were not just about winning a tennis match or getting a date—they were the key to personal, political, and religious success.

ANCIENT CHINA MEETS MODERN SCIENCE

Chinese thinkers living more than two thousand years ago inhabited a social and religious world very different from our own. I therefore find it particularly revealing that over the past few decades many of their insights about the importance of spontaneity have been rediscovered by contemporary science. A growing literature in psychology and neuroscience suggests that these thinkers had a much more accurate picture of how people really think and behave than we find in recent Western philosophy or religious thought and that early Chinese debates about how to attain *wu-wei* reflect real tensions built into the human brain. Scientists are beginning to better appreciate the role that "fast and frugal" unconscious thinking plays in everyday human life and now have a clearer sense of why spontaneity and effectiveness hang together.

We also now know something about the psychological mechanisms involved in *wu-wei*, including some details about which brain regions are being turned off, as it were, and which are firing at full speed. Technologies such as fMRI (functional magnetic resonance imaging), which tracks neural activity by measuring blood flow to the brain, even allow us to see vivid images of the *wu-wei* brain in action. This is not to fetishize brain science as some magic pathway to the truth. The snazzy fMRI diagrams that festoon popular articles and books are not snapshots of the mind itself. That said, they are almost certainly telling us *something*, ideally something we didn't already know about how the brain works. They are a helpful piece in the puzzle. Moreover, contemporary neuroscience is useful because it moves us toward a greater appreciation of the complexities of our embodied mind. When we are consciously struggling with temptation, for instance, what's involved is not a disembodied soul pulling on the reins of a brutish body, but a set of brain regions dedicated to certain functions in conflict with another set of brain regions. Neuroscience therefore gives us a more accurate description of ourselves than we've ever had.

Another important piece in the spontaneity puzzle comes from evolutionary psychology, which gives us insight into why *wu-wei* is so pleasant for the individual and attractive to others. Things tend to give pleasure when evolution approves of them: think orgasms or chocolate. It feels good to be in *wu-wei* because a whole slew of tasks simply can't be performed by our plodding, conscious minds—we need to unleash the power of our fast, unconscious processes in order to get them done. Moreover, we are attracted to people in *wu-wei* because we *trust* the automatic, unconscious mind. We have a very strong intuition—increasingly confirmed by work in cognitive science—that the conscious, verbal mind is often a sneaky, conniving liar, whereas spontaneous, unselfconscious gestures are reliable indicators of what's really going on inside another person. Physiologically, it is hard to consciously kick-start spontaneity, and this is probably why we value it so much in our social lives. It's also why we find spontaneous people attractive and trustworthy, as well as why conscious attempts to simulate spontaneous ease—think of any of the commercially marketed dating strategies, from "the Rules" to "the System"—tend to fall flat.

So although early Chinese thinkers had all sorts of metaphysical theories about how and why *de* is attractive to others, it can probably be explained by this very simple psychological fact: spontaneous behavior is hard to fake, which means that spontaneous, unselfconscious people are unlikely to be fakers. We're also attracted to effectiveness, and people in *wu-wei* tend to be socially competent as they move through life. Taken together, these considerations give us an empirically grounded, scientific basis for taking both *wu-wei* and *de* seriously as concepts that can help us make sense of our own lives. We'll therefore spend time exploring the deep evolutionary and neuroscientific reasons for why *wu-wei* is effective, why *de* works, and why certain contemporary ideas of spontaneity miss important aspects of the *wu-wei* experience.

CONTEMPORARY INSIGHTS FROM
ANCIENT PHILOSOPHY

If modern science can tell us so much about *wu-wei*, why bother with these early Chinese guys? (And, yes, as far as we know they were all guys.) As a historian of early Chinese thought—and, not incidentally, someone raised in New Jersey, with a Jerseyite's low tolerance for B.S.—I get a psychosomatic headache from people who glorify "the East" as if it were an exclusive and unfailing source of spiritual wisdom. Nonetheless, there is a kernel of truth lurking at the center of New Age exoticism. There are several important ways that early Chinese thought can help us get beyond some of our philosophical and political hang-ups and better prepare us to grasp the biological and cultural worlds we inhabit.

If we take "the West" to refer to the dominant mode of thinking in post-Enlightenment Europe and its colonies, one thing we *can* say is that it tends to portray rational thought as the essence of human nature, and reasoning as something that occurs in an ethereal realm completely disconnected from the noise and heat of the physical world around us. This view is strongly *dualistic* in the sense that the mind, and its supposedly abstract rationality, are seen as radically distinct from, and superior to, the body and its emotions. Although some sort of mind-body dualism seems to be a human psychological universal, the tradition that can be traced from Plato down to Descartes converted vague intuitions about the distinction between *people* (who possess minds) and *things* (which do not) into a bizarre metaphysical dichotomy between a completely invisible, disembodied *mind* and the physical *stuff* that makes up our material world.

This strongly dualist perspective has not only confused us about ourselves but also had an extraordinarily bad effect on science. Early (mid-twentieth-century) cognitive scientists treated the human mind as a brain in a vat, performing abstract information processing, and this led them down some very unproductive paths. Fortunately, over the past few decades cognitive science has begun to free itself from the conceptual shackles of dualism and to treat human thought as fundamentally "embodied." What this means is that our thinking is grounded in concrete experiences and that even what seem like quite abstract concepts are linked to our bodily experience through analogy and metaphor. It is difficult to think about "justice" without summoning the image of a scale or some physical thing being evenly split; when reasoning about our lives, we inevitably think in terms of journeys and paths not taken. The embodied view of cognition also views thought as inherently tied to feeling, which calls into question any rigid distinction between rationality and emotion. Moreover, cognitive scientists are beginning to emphasize the fact that the human brain is designed primarily for guiding *action*, not for representing abstract information—although it can also do this when necessary. This "embodied cognition" revolution was at least partially inspired by insights from Asian religious thought—including both early Chinese and later Chinese Buddhist accounts of *wu-wei*—which makes this book's melding of cognitive science and Chinese thought particularly relevant.

Although typically a bit slower off the starting block than scientists, Western philosophers are also beginning to realize the importance of both empirical knowledge and alternative traditions for their discipline. A small but growing number of psychologically attuned philosophers now recognize that the early Chinese tradition, with its embodied model of the self, offers an important corrective to the tendency of modern Western philosophy to focus on conscious thought, rationality, and willpower. For instance, while recent Western thought has emphasized the importance of abstract, representational knowledge—that is, information about the world, like the fact that Rome is the capital of Italy or that $e = mc^2$—early Chinese thought instead emphasized what we could call *know-how*: the practical, tacit, and often unformulizable ability to *do* something well. I cannot explain exactly how to ride a bike, but I can ride one.

In fact, as we'll discuss later, consciously focusing on how to ride, or trying to explain the process in words to others, may actually impair your ability to do it.

For the early Chinese thinkers whom we'll meet in this book, the culmination of knowledge is understood, not in terms of grasping a set of abstract principles, but rather as entering a state of *wu-wei*. The goal is to acquire the ability to move through the physical and social world in a manner that is completely spontaneous and yet fully in harmony with the proper order of the natural and human worlds (the *Dao* or "Way"). Because of this focus on knowing *how* rather than knowing *this* or *that*, the Chinese tradition has spent a great deal of energy over the past two thousand years exploring the interior, psychological feel of *wu-wei*, worrying about the paradox at the heart of it, and developing a variety of behavioral techniques to get around it. The ideal person in early China is more like a well-trained athlete or cultivated artist than a dispassionate cost-benefit analyzer. This better fits both our intuitive sense of what real human excellence looks like and our best current scientific understanding of how the mind works.

In addition to helping us get beyond strong mind-body dualism, the Chinese concepts of *wu-wei* and *de* reveal important aspects of spontaneity and human cooperation that have slipped through the nets of modern science, which is still very grounded in another basic feature of Western thought: extreme individualism. The ideal person in Western philosophy is not only disembodied but also radically alone. For the past couple hundred years in the West, the dominant view of human nature has been that we are all individual agents pursuing our own self-interest, responsive only to objective rewards and punishments. Human societies, according to this narrative, were formed when lone hunter-gatherers—typically men, suspiciously unaccompanied by their mates, children, elderly parents, or sick friends—met in a clearing, negotiated the rules they would live under, and then shook hands and agreed to cooperate. As economists and political scientists have only recently begun to realize, this is a fairy tale cooked up over the last century or two by a bunch of elite, landowning males—what the philosopher Annette Baier has scathingly referred to as "a collection of clerics, misogynists, and puritan bachelors."

In reality, we are not autonomous, self-sufficient, purely rational individuals but emotional pack animals, intimately dependent on other human beings at every stage of our lives. We get along, not because we're good at calculating costs and benefits, but because we are emotionally bound to our immediate family and friends and have been trained to adopt a set of *values* that allows us to cooperate spontaneously with others in our society. These shared values are the glue that holds together large-scale human groups, and a key feature of these values is that they need to be embraced sincerely and spontaneously—in an *wu-wei* fashion—to do their job. This is why the tensions surrounding *wu-wei* and *de* are linked to basic puzzles surrounding human cooperation, especially in the anonymous, large communities we tend to inhabit today.

Moreover, situating *wu-wei* in its original, early Chinese context helps us to see how it is a fundamentally spiritual or religious concept. One of the key features of the *wu-wei* state is a sense of being absorbed in some larger, valued whole—typically referred to as the *Dao* or "Way." While modern readers are highly unlikely to subscribe to the early Chinese religious worldview, I'll show that, even for us, something quite similar to the Way is at work in the background of any genuine *wu-wei* experience: a sense of being at home in some framework of values, however vague or tenuous. This will make it clear how *wu-wei* differs from modern psychological concepts such as "flow," allowing us to recover the crucial *social* dimension of spontaneity.

There is one final benefit to looking at *wu-wei* and *de* in their original Chinese context, at least for readers outside East Asia. It's important to realize that Chinese culture has never entirely lost its focus on *wu-wei*—it never took that trip down the rabbit hole of hyper-rationality and extreme individualism. The thinkers explored in this book are still alive and well in the contemporary Chinese mind-set. Because Chinese is not a phonetic language, its written form has remained essentially the same over several millennia, even as the spoken dialects have undergone massive changes. The texts that we will be exploring were written from roughly the fifth to the third century B.C.E., in so-called "classical Chinese," which continued to serve as the sole literary language of China until the early twentieth century. Indeed, until quite recently, classical Chinese was the lingua franca and medium of scholarship across East Asia. For much of this history, the texts that we will be discussing were memorized by every educated person throughout the Chinese cultural sphere. In terms of sheer numbers, then, the *Analects* of Confucius has influenced far more people than the Bible. Even today large portions of these texts are internalized by schoolchildren throughout East Asia, and the classical language—bearing with it particular ways of thinking—permeates all of the spoken dialects of China.

The result is an unusually high degree of continuity between the formative period of Chinese thought and the modern culture of China in particular, and East Asia more generally. To take one example, the contemporary East Asian emphasis on personal relationships (*guanxi*) and informal business networks initially strikes Westerners doing business in China as simply a front for corruption and nepotism—and no doubt it often is. However, it has its own rationale, one grounded in an ancient Confucian preference for character-based, intuitive judgments over rigid rule following. You deal with people face to face because their spontaneous expressions and casual remarks—their *wu-wei* behavior, the quality of the *de* that they emanate—tell you everything that you need to know about them. The best way to ensure that agreements are honored is to forge intense personal loyalties, typically over the course of several evenings fueled by good food and staggering quantities of alcohol. Even in ancient China there was a clear awareness that such personal networks are open to abuse, and there were many who argued that the only way to prevent this was to replace spontaneous, trust-based social ties with rule-bound, impersonal public institutions. In practice, China ended up with a mixture of the two, but the culturally dominant social norms have continued to be grounded in *wu-wei* and *de*. One cannot understand contemporary China without grasping this point. So, among other things, this book is intended to provide some insights into the nature of Chinese thought, which should be helpful for anyone living in an age where China has reemerged—after what was really just a short hiatus—as a major world power.

REDISCOVERING THE VALUE OF SPONTANEITY

At a very broad level, early Chinese accounts of spontaneity provide us with a unique window into an aspect of human spiritual and social life that has been overlooked in recent centuries in the West. More concretely, although this book will not present a 100 percent guaranteed, ten-step program for achieving *wu-wei*, you will end up learning quite a bit about strategies that have worked for people in the past. One reason that no one has ever come up with a single, surefire technique for achieving *wu-wei* is that the barriers to spontaneity vary from person to person and situation to situation. This means that having a grab bag of strategies at one's disposal is probably quite useful. Whatever technique works in helping you fall asleep the night before a big meeting may prove useless when it comes to helping you get into the zone during your tennis match that afternoon. What tends to help me—an introverted, slow-moving academic—to relax into *wu-wei* at the beginning of a workday (silence, solitude, sunlight, and large quantities of coffee) may be completely different from what an extroverted, hyperactive actor needs (a stiff shot of bourbon, loud music, and a frenetic group brainstorming session). My hope is that by exploring the distilled wisdom of ancient traditions of thought—as well as the best findings from contemporary cognitive science—you will gain new insights that you can apply to your own life.

Moreover, simply adopting *wu-wei* and *de* into our vocabularies forces us to fundamentally change the way we think about human behavior and relationships. We tend to think of spontaneity—when we do at all—as a helpful ingredient at a cocktail party or as something that concerns only artists or athletes. There is increasing evidence, however, that the early Chinese were right that spontaneity is, in fact, a cornerstone of individual well-being and human sociality. This means that the paradox of *wu-wei* is also more central than we realize, and the problem of overcoming it more urgent than we think.

We have been taught to believe that the best way to achieve our goals is to reason about them carefully and strive consciously to reach them. Unfortunately, in many areas of life this is terrible advice. Many desirable states—happiness, attractiveness, spontaneity—are best pursued indirectly, and conscious thought and effortful striving can actually interfere with their attainment. In the pages that follow, we'll learn how to foster spontaneity in our own lives and gain some insight into how spontaneity affects our relationships with others. Our exploration of *wu-wei* and *de* will offer no easy solutions, no corny "ancient Chinese secrets" that will instantly turn us into serene Zen masters. I am convinced, though, that becoming aware of the power and problems of spontaneity—and thinking through these issues with both early Chinese thinkers and cognitive scientists at our side—can give us a deeper understanding of how we move through the world and interact with others, and can help us to do it more effectively.

1

Skillful Butchers and Graceful Gentlemen

THE CONCEPT OF *WU-WEI*

THE STORY OF BUTCHER DING IS PERHAPS THE BEST-KNOWN AND most vivid portrayal of *wu-wei* in the early Chinese tradition. The butcher has been called upon to play his part in a traditional religious ceremony involving the sacrifice of an ox, in a public space with the ruler and a large crowd looking on. This is a major religious event, and Butcher Ding is at center stage. The text is not specific, but we are probably witnessing a ceremony to consecrate a newly cast bronze bell. In this ritual, the still-smoking metal is brought fresh from the foundry and cooled with the blood of a sacrificial animal—a procedure that demands precise timing and perfectly smooth execution.

Butcher Ding is up to the task, dismembering the massive animal with effortless grace: "At every touch of his hand, every bending of his shoulder, every step of his feet, every thrust of his knee—swish! swoosh! He guided his blade along with a whoosh, and all was in perfect tune: one moment as if he were joining in the Dance of the Mulberry Grove, another as if he were performing in the Jingshou Symphony." The Dance of the Mulberry Grove and the Jingshou Symphony were ancient, venerated art forms: Ding's body and blade move in such perfect harmony that a seemingly mundane task is turned into an artistic performance. Lord Wenhui is amazed and is moved to exclaim, "Ah! How wonderful! Can skill really reach such heights?" Butcher Ding puts down his cleaver and replies, "What I, your humble servant, care about is the Way [Dao, 道], which goes beyond mere skill." He then launches into an explanation of what it feels like to perform in such a state of perfect ease:

> When I first began cutting up oxen, all I could see was the ox itself. After three years, I no longer saw the ox as a whole. And now— now I meet it with my spirit and don't look with my eyes. My senses and conscious awareness have shut down and my spiritual desires take me away. I follow the Heavenly pattern of the ox, thrusting into the big hollows, guiding the knife through the big openings, and adapting my motions to the fixed structure of the ox. In this way, I never touch the smallest ligament or tendon, much less a main joint.

The result is that Butcher Ding is not so much cutting up the ox as *releasing* its constituent parts, letting the razor-sharp edge of his cleaver move through the spaces between the bones and ligaments without encountering the slightest resistance:

> A skilled butcher has to change his cleaver once a year, because he cuts; an ordinary butcher has to change his cleaver once a month, because he hacks. As for me, I have been using this particular cleaver for nineteen years now, and have cut up thousands of oxen with it, and yet its edge is still as sharp as when it first came off the whetstone. Between the joints of the ox there is space, and the edge of the blade has no thickness; if you use that which has no thickness to pass through gaps where there is space, it's no problem, there's plenty of room to let your cleaver play. That's why, after nineteen years, the edge of my blade looks like it just came from the whetstone.

It is not *all* smooth sailing. Occasionally Butcher Ding's effortless dance is interrupted when he senses trouble, at which point his conscious mind seems to reengage a bit, although he still remains completely relaxed and open to the situation confronting him: "Whenever I come to a knot, I see the difficulty ahead, become careful and alert, focus my vision, slow my movements, and move the blade with the greatest subtlety, so that the ox simply falls apart, like a clod of earth falling to the ground." Lord Wenhui clearly sees something in this account that goes far beyond simply cutting up oxen. "Wonderful!" he exclaims. "From the words of Butcher Ding I've learned how to live my life!" This remark signals to us that we should be taking the story of the ox as a metaphor: we are Butcher Ding's blade, and the bones and ligaments of the ox are the barriers and obstacles that we face in life. Just as Butcher Ding's blade remains razor-sharp because it never touches a bone or ligament—moving only through the gaps in between—so does the *wu-wei* person move only through the open spaces in life, avoiding the difficulties that damage one's spirit and wear out one's body. This is a metaphor that has not lost any of its power. I, for one, can attest that, after forty-odd years of sometimes hard living, my own blade feels a bit nicked and dull.

Another of my favorite portrayals of *wu-wei* also concerns an artisan. A woodcarver named Qing has received commissions to carve massive wooden stands for sets of bronze bells—precisely the sort of bells that were consecrated in Butcher Ding's ritual sacrifice. Again, this is high-stakes public art, commissioned by the ruler himself, and involving the promise of a juicy monetary reward and official honors. As with Ding, Qing demonstrates almost supernatural skill: the bell stands that he produces are so exquisite that people think they must be the work of ghosts or spirits. Like Butcher Ding, he is praised by his ruler, who exclaims, "What technique allows you to produce something that beautiful?" Again, like Ding, the woodcarver demurs, denying that what he does is all that special. "I, your servant, am merely a humble artisan. What technique could

I possibly possess?" After being pressed a bit, though, he acknowl-
edges that perhaps there is a secret to his success, having to do with
how he prepares himself mentally to begin the work: "When I am
getting ready to make a bell stand, the most important thing is not
to exhaust my energy [*qi*], so first I fast in order to still my mind.
After I have fasted for three days, concerns about congratulations or
praise, titles or stipends no longer trouble my mind. After five days,
thoughts of blame or acclaim, skill or clumsiness have also left my
mind. Finally, after fasting for seven days, I am so completely still
that I forget that I have four limbs and a body." The idea of carving
a bell stand without a sense of one's limbs or body might seem odd,
but the point is that Qing has so focused his attention that all external considerations have fallen away. "There is no more ruler or
court," he explains, "my skill is concentrated and all outside distractions disappear." He's ready to get to work.

> Now I set off for the mountain forest to observe, one by one, the
> Heavenly nature of the trees. If I come across a tree of perfect
> shape and form, then I am able to see the completed bell stand al-
> ready in it: all I have to do is apply my hand to the job and it's done.
> If a particular tree does not call to me, I simply move on. All that I
> am doing is allowing the Heavenly within me to match up with the
> Heavenly in the world—this is probably why people mistake my
> art for the work of the spirits!

It's striking how similar this story is to the lore surrounding a great
public artist from an entirely different time and culture, Michelangelo. When questioned about his own apparently supernatural
sculpting talents, he supposedly replied that, when given a commis-
sion, he simply waited until he found a piece of marble in which he
could already see the sculpture. All he then had to do was cut away
the stone that didn't belong. Here, as with Woodcarver Qing, there
is a sense that the materials themselves dictate the artistic process.

The artist's own contribution is portrayed as minimal, and the creative act is experienced as completely effortless.

The stories of Butcher Ding and Woodcarver Qing both come from a book called the *Zhuangzi*, one of the two Daoist works that we will be looking at, and the richest hunting ground for *wu-wei* stories among Warring States texts. Characterizations of *wu-wei* in the other of our early Daoist texts, the *Laozi*, take the form of concise, cryptic poems rather than stories—much of the book probably rhymed in the original Chinese pronunciation, which we can now only imprecisely reconstruct. A typically mysterious passage from the *Laozi* describing the "Way of Heaven" is clearly meant to provide a model for how a properly cultivated person should move through the world:

> The Way of Heaven
> Excels in overcoming, though it does not contend;
> In responding, though it does not speak;
> In spontaneously attracting, though it does not summon;
> In planning for the future, though it is always relaxed.
> The Net of Heaven covers all;
> Although its mesh is wide, nothing ever slips through.

The "wide mesh" that nonetheless captures everything is reminiscent of the relaxed concentration of Butcher Ding or Woodcarver Qing: at ease and yet open, profoundly attuned to the environment. Unlike our Zhuangzian exemplars, however, who attain perfection only after long periods of training in particular skills, the Laozian sage attains *wu-wei* by *not* trying, by simply relaxing into some sort of preexisting harmony with nature:

> Do not go out the door, and so understand the whole world;
> Do not look out the window, and understand the Way of Heaven.
> The farther you go, the less you know.

This is why the sage understands the world without going abroad,
Achieves clarity without having to look,
And attains success without trying.

These sorts of passages, where *wu-wei* is an explicit focus, are quite common throughout the *Zhuangzi* and the *Laozi*, which is why the concept of *wu-wei* is typically associated with Daoism.

What is less widely appreciated, however, is that the sort of effortless ease and unselfconsciousness that characterizes these Daoist accounts also plays a central role in early Confucianism. This may come as a surprise, because Confucianism is typically associated with hidebound traditionalism and stuffy ritual—both of which strike us as the opposite of *wu-wei*. It can't be denied that the Confucians do a lot to earn this reputation. In the early stages of training, an aspiring Confucian gentleman needs to memorize entire shelves of archaic texts, learn the precise angle at which to bow, and learn the length of the steps with which he is to enter a room. His sitting mat must always be perfectly straight. All of this rigor and restraint, however, is ultimately aimed at producing a cultivated, but nonetheless genuine, form of spontaneity. Indeed, the process of training is not considered complete until the individual has passed completely beyond the need for thought or effort.

Confucius himself, in a passage that serves as a wonderfully concise spiritual autobiography, portrays *wu-wei* as the goal for which he has spent his entire life striving: "The Master said, 'At fifteen I set my mind upon learning; at thirty I took my place in society; at forty I became free of doubts; at fifty I understood Heaven's Mandate; at sixty my ear was attuned; and at seventy I could follow my heart's desires without transgressing the bounds of propriety.'" The phrase "my ear was attuned" literally means "my ear flowed along / went with the flow" and suggests that when hearing the teachings of the ancients Confucius immediately grasped and took joy in them. By age seventy, he had so internalized the Confucian Way that he could act upon whatever thought or desire popped into his head and yet still behave in a perfectly moral and exemplary fashion. The end result looks as effortless and unselfconscious as that of the Zhuangzian butcher or Laozian sage but is, in fact, the product of a lifelong process of training in traditional cultural forms.

Confucius's form of *wu-wei*—an effortless, unselfconscious, but eminently *cultured* spontaneity—was inherited as an ideal by his two Warring States followers, Mencius and Xunzi, although they disagreed profoundly about what's required to reach this state. Mencius tried to split the difference, as it were, between the Daoists and Confucius by presenting *wu-wei* as the natural outgrowth of cultivating our nature. For him, morally proper *wu-wei* was like a sprout waiting to break through the ground, or a body prepared to move with a catchy beat. Xunzi, on the other hand, was unimpressed by the Daoist celebration of nature and returned to the model championed by Confucius, whereby *wu-wei* was the result of a lifetime of rigorous education. For Xunzi, "not trying" was neither easy nor fun: the perfection of form and emotion that finds its ideal expression in dance was, for him, a hard-won achievement resulting from years of difficult training and cultural learning. In any case, this preoccupation with how to cultivate *wu-wei* was at the center of early Chinese controversies about how to attain the good life. This is a conversation worth paying attention to, because it brings to the forefront ideas, like spontaneity and charisma, that have fallen through the cracks of our contemporary mind-set.

YOUR BRAIN ON *WU-WEI*

In the early Chinese accounts of *wu-wei* described above, a couple of features are immediately apparent. First, although there is only one Butcher Ding or Confucius in the world, these *wu-wei* exemplars experience themselves as *split*. They seem to feel a gap between an "I" (the locus of consciousness and personal identity) and various forces—spiritual desires, desires of the heart—that take over when they enter *wu-wei*. *Wu-wei* is characterized by an internal sense of effortlessness and unselfconsciousness, even though the person in *wu-wei* may actually be very active in the world. Someone or something else must be doing the work besides the conscious mind that we normally think of as "us." Second, people in *wu-wei* are extremely effective: huge oxen fall apart with a few swipes of the blade, and complex social situations are negotiated with masterly aplomb. My guess is that we have all experienced this combination of effortlessness and effectiveness at some point in our lives. While we are completely absorbed in chopping and sautéing, a complex dinner simply assembles itself before our eyes. Fully relaxed, we breeze through an important job interview without even noticing how well it's going. Our own experiences of the pleasure and power of spontaneity explain why these early Chinese stories are so appealing and also suggest that these thinkers were on to something important. Combining Chinese insights and modern science, we are now in a position to understand how such states can actually come about.

Colloquially, we often speak of ourselves as if we were split in two: "I couldn't make *myself* get out of bed this morning," "I had to force *myself* to be calm," "I had to hold *my tongue*." Although we use such phrases all the time, if you think about them they're a bit weird. Who is the self who doesn't want to get out of bed, and what is its relationship to me? Does my tongue really have a will of its own, and how do I go about holding it? (And who am I if not my tongue?) Since there is always only one "me" involved, this split-self talk is clearly metaphorical rather than literal. At the same time, the fact that we fall back upon this kind of language so frequently means that it must reflect something important about our experience. And talk of split selves is certainly not limited to English: we can see it in many *wu-wei* stories from early China that involve a narrative "I" confronting a part of the self that is more or less autonomous.

A typically colorful example from the *Zhuangzi* begins by introducing us to a *kui* (*kway*, 夔), a strange mythical creature that gets around by slowly and painfully hopping on a single foot. Glancing enviously at a millipede tearing past at full speed, its thousands of little legs busily churning along, the *kui* says, "The only way I have of getting around is to hop along on this one leg of mine, and although I have only a single leg, I still can't manage to control it very well. You, on the other hand, have *thousands* of legs—how in the world do you control them all?" The millipede replies, "You don't understand, I don't control them at all. It's like spitting on the ground—you just up and do it. . . . All I do is put my Heavenly mechanism in motion, and it does all the work, I have no idea how."

Only the author of the *Zhuangzi* would be irreverent enough to use spitting as an example of our "Heavenly mechanism" in action. It certainly lacks the romanticism or mystical flair that is often associated with Daoism. But this may be precisely the point: when we're in *wu-wei*, it doesn't feel like anything special, it just *is*. Spitting is simply one example of a massive number of things our body knows how to do that we'd be hard-pressed to consciously control or explain in words. Walking with our own two legs, rather than the millipede's thousands or the *kui* monster's one, is another good example. We don't worry about how to walk, we don't consciously monitor ourselves while we're walking, we just walk. In fact, actually *thinking* about walking while trying to walk is a great way to trip. We'll talk in more detail later about the deleterious effects of this kind of conscious monitoring, but for now let's just note that there are many things our body knows how to do without any input at all from our conscious mind, and when we reflect upon this we get a strong sense of being split between a conscious "I" and an unconscious body, which often seems to have a mind of its own.

Recent research suggests that there might be some basis to this idea. Although there is only one me, in an important functional sense we *are* divided into two beings. There is now general agreement that human thought is characterized by two distinct systems that have very different characteristics. The first and most important of these (tacit, hot cognition, or "System 1") is fast, automatic, effortless, and mostly unconscious, corresponding roughly to what we think of as "the body" and what Zhuangzi calls the "Heavenly mechanism." The second (explicit, cold cognition, or "System 2") is slow, deliberate, effortful, and conscious, corresponding roughly to our "mind"—that is, our conscious, verbal selves.

So if I say that I had to force myself not to reach for that second helping of tiramisu, there is a more than metaphorical struggle going on. My conscious, cold system, which is concerned about long-range issues like health and weight gain, is fighting to control the more instinctive hot system, which really likes tiramisu and doesn't share my cold system's concerns about the consequences. This isn't because hot cognition doesn't take future consequences into account. The problem is that this system's conception of relevant consequences was fixed a long time ago, evolutionarily speaking, and is fairly rigid. "Sugar and fat: *good*" was for most of our evolutionary history a great principle to live by, since acquiring adequate nutrition was a constant challenge. For those of us fortunate enough to live in the affluent industrialized world, however, sugar and fat are so widely and freely available that they no longer represent unqualified goods—on the contrary, allowing ourselves to indulge in them to excess has a variety of negative consequences. The great advantage of cold cognition is that it is capable of changing its priorities in light of new information. So another way to think about how the systems differ is that hot cognition is evolutionarily older and more rigid, while cold cognition is evolutionarily newer and more flexible—and therefore more likely to adapt to novel behavioral consequences.

These two systems are even, to some degree, neuroanatomically distinct—that is, they are implemented in different parts of the brain. In fact, our first hint that the two systems existed came from clinical cases where selective brain damage allowed researchers to watch one of them functioning without the other. Anyone who has seen the movie *Memento* (2000) is familiar with a condition called anterograde amnesia: patients afflicted with this condition cannot form new, explicit short-term memories. They remember who they are and the more distant past but are condemned—consciously, at least—to a perpetual forgetting of the present. What's interesting is that, although these patients can't form new conscious memories, at a subconscious level they are able to form new, *implicit* ones. They cannot consciously recall ever having met the doctor who greets them every day with a thumbtack hidden in his palm, but for some reason they don't want to shake his hand. We see a similar disjunction when it comes to different types of skills: unconscious "knowing how" seems distinct from conscious "knowing that." As with emotional memories, the two types of knowledge seem to be created and preserved in different parts of the brain. Not only can amnesiac patients "remember" not to shake hands with Dr. Thumbtack, they are also able to pick up new physical skills after a period of instruction without having any conscious memory of the training, even while remaining completely unable to explain how or why they have this new ability.

So although talk of "mind" and "body" is technically inaccurate, it does capture an important functional difference between two systems: a slow, cold, conscious mind and a fast, hot, unconscious set of bodily instincts, hunches, and skills. "We" tend to identify with the cold, slow system because it is the seat of our conscious awareness and our sense of self. Beneath this conscious self, though, is another self—much bigger and more powerful—that we have no direct access to. It is this deeper, more evolutionarily ancient part of us that knows how to spit and move our legs around. It's also the part that we are struggling with when we try to resist that tiramisu or drag ourselves out of bed for an important meeting.

The goal of *wu-wei* is to get these two selves working together smoothly and effectively. For a person in *wu-wei*, the mind is embodied and the body is mindful; the two systems—hot and cold, fast and slow—are completely integrated. The result is an intelligent spontaneity that is perfectly calibrated to the environment. The fluidity with which Butcher Ding's cleaver glides through the ox is suggested in the story by onomatopoetic descriptors—*swish, swoosh, whoosh*—that are a hallmark of the *Zhuangzi* and a real headache for translators. The ease perceived by observers of his performance mirrors Ding's own internal experience, in which his "spiritual desires" take him away and the ox falls apart effortlessly. Similarly, Woodcarver Qing describes his artistic process as simply letting the bell stand reveal itself to him. All he needs to do is apply his hand and the bell stand emerges, as if by magic.

How this process of integration works is suggested in stories like that of Butcher Ding. Recall the three stages of "seeing" that the butcher describes to his lord: "When I first began cutting up oxen, all I could see was the ox itself. After three years, I no longer saw the ox as a whole. And now—now I meet it with my spirit and don't look with my eyes." These lines appear to describe the act of "seeing" by using different parts of the self. When Butcher Ding is a complete novice, and all he can see is "the ox itself," he is merely looking with his eyes, gazing upon this massive, daunting creature that he must somehow reduce to pieces. Anyone who has ever seen an ox up close—not a common experience in our modern world—can vividly imagine the plight of the neophyte Butcher Ding. There he is, standing in front of this massive wall of hair and flesh, cleaver in hand, with no idea where to begin or what to expect once he makes the first cut. I, for one, would look for another line of work.

We have to believe, though, that Butcher Ding showed more persistence, and after three years of practice and training he has reached a point where he "no longer sees the ox as a whole." Perhaps Butcher Ding now looks at the ox and sees, superimposed upon it, something like the charts that hang in butchers' shops, illustrating the different cuts of meat. The ox, for him, is no longer simply a dumb, inert object in his path. Drawing upon his training and analytic mind, Butcher Ding now perceives it in terms of its constituent parts, as a set of cuts that he can make, or a set of challenges to be traversed. At this point he sees the ox the way a grand master sees a chessboard in midplay: where we might see only a checkered board with wooden pieces on it, a chess master sees lines of force, zones of danger, and paths of opportunity.

Finally, Butcher Ding reaches a stage where, as he puts it, he no longer looks with his eyes: "My senses and conscious awareness have shut down," he explains, "and my spiritual desires take me away." I've always explained this stage of perception to my students by way of a *Star Wars* analogy. If you are of a certain age, you will vividly remember the finale of the first *Star Wars* movie (1977). Luke Skywalker has embarked upon a daring, long-shot mission to destroy the massive, almost impervious Death Star. The only way for Luke to circumvent the Death Star's defenses is to fly down a narrow trench in its surface and score a direct missile hit at a weak point in its armor. Being pursued by Darth Vader and his minions, Luke has only one chance to get it right. As he approaches his destination and begins to activate his targeting computer, he hears in his mind the voice of the recently slain Obi-Wan (in that wonderfully sonorous Alec Guinness voice): *Use the Force, Luke . . . Let go!* To the consternation of those tracking him at the rebel base, Luke shuts down his computer, closes his eyes, and taps into the Force in order to *feel* the right time to release his missiles. Of course he scores a direct hit, the Death Star is destroyed, and he is reunited as a hero with that universal object of late seventies/early eighties adolescent boy desire, Princess Leia. (I still find that bun hairstyle incredibly erotic.)

If Zhuangzi were still alive and suitably lawyered up, I'd advise him to go after some of the royalties from this movie, because Luke's strategy is essentially that of the Butcher Ding: "My senses and conscious awareness have shut down" (I've shut off my targeting computer) and "my spiritual desires take me away" (I let the Force guide me). In fact, the Zhuangzi–*Star Wars* connection is a historically accurate one. George Lucas, in conceptualizing the *Star Wars* mythology—particularly the Jedi Knights and figures such as Yoda—was at least partly inspired by the mystical *bushido* ("Way of the Warrior") code of the Japanese samurai, which combined a bit of Confucianism with big doses of Japanese Zen Buddhism. In turn, Japanese Zen Buddhism is derived from Chinese Chan Buddhism, and Chan Buddhism is (as I like to say when I want to annoy my

Buddhologist colleagues) basically just Zhuangzi in Buddhist drag. Dress up Butcher Ding in a samurai outfit, swap out his cleaver for a sword, and he would not be at all out of place in *The Last Samurai*. So there is a direct line of intellectual descent between the story of Butcher Ding in the *Zhuangzi* and that of Luke Skywalker in *Star Wars*. The mythology that Lucas eventually ended up with is a bit of a mishmash—the whole idea of there being a "dark side" to the Force comes out of Christian notions about evil and wouldn't make sense in an East Asian religious context—but the image of a Jedi Knight being moved solely by the Force looks very much like Butcher Ding closing his eyes and allowing himself to be guided by his "spiritual desires" as he dances his way through the ox.

There is no evidence, as far as I know, that George Lucas is a fan of the *Analects* of Confucius, but we can see the same combination of effortlessness and unselfconsciousness in all of the early Chinese exemplars. There is a great passage in the *Analects* describing the utter rapture that seized Confucius when he first heard the music of the ancient sage kings who served as his models of virtue: "For three months after, he did not even notice the taste of meat. He said, 'I never imagined that music could be so sublime.'" It is this joy that makes Confucius indifferent to material wealth and reputation and allows him to give free rein to his every desire without ever doing anything improper. We see something similar in the story of Woodcarver Qing, where his ability to perceive in a glance the tree that contains his bell stand is predicated on attaining a state of complete unselfconsciousness. More specifically, he needs to clear his mind of all external considerations, such as money, fame, honor, and a sense of his own physical self.

How are we to understand this from a contemporary perspective? Can we do without mystical entities like "spiritual desires" or the Force and still make sense of the experiences being described here? I think we can. To do so, we need a clear understanding of what *effort* and *consciousness* feel like from the inside.

Let's begin with a little exercise. Go down the column of words below, and as quickly as possible read each word silently and then say out loud either "upper" or "lower," depending on whether the word itself is written in upper- or lowercase letters.

UPPER
lower
lower
upper
LOWER

Unless you are an alien cyborg from Alpha Centauri, you were probably cruising along until you reached the last two, where you stumbled a bit and took longer to say "lower" while reading *upper*, and then "upper" while reading *LOWER*. That slight catch as you began to talk—that feeling of needing to stop yourself, to not *read* the word but instead focus on its case—has been referred to as the *oomph* that is the hallmark of conscious will or effort. A task that presents someone with this kind of mismatch between the meaning of a word and its physical appearance is commonly referred to as a "Stroop task," after the English psychologist who published a paper on the effect in the 1930s, originally using words printed in incompatible colors (for example, the word *green* printed in red ink). The Stroop task is a classic example of what's called a *cognitive control* or *executive control* task—that is, a situation where the cold, conscious mind (System 2) has to step in and override automatic, effortless processes (System 1).

Brain imaging studies suggest that a couple brain areas in particular are involved in cognitive control: the *anterior cingulate cortex* (ACC) and the *lateral prefrontal cortex* (lateral PFC). We'll be referring to these together as the "cognitive control regions" of the brain. There is still some debate about the precise role played by each of these regions, but one plausible characterization is that the ACC is a kind of smoke detector, and the lateral PFC is the fire response team. Like a smoke detector, the ACC is in constant monitoring mode, waiting to detect a whiff of danger, such as an instance of cognitive conflict. In the case of the Stroop task, we've got two automatic processes that are in conflict: the identification of a typeface or color versus the automatic processing of a simple word (assuming you're literate and it's your native language). This conflict alerts the ACC, which then sends out an alarm to the lateral PFC to come deal with the situation.

The lateral PFC is responsible for many higher cognitive functions, such as the integration of conscious and unconscious knowledge, working memory (the small spotlight of consciousness that allows us to focus on explicit information), and conscious planning. Most relevantly, when it comes to the case of the Stroop task, the lateral PFC also exerts control over other areas of the brain by strengthening the activation of task-relevant networks at the expense of other networks. By weakening certain neural pathways, the lateral PFC essentially tells them to stop doing what they are doing, which is the neural equivalent of fire-retarding foam.

In the Stroop task presented above, you were asked to read the word *LOWER* but say the word *upper*. The ACC lets the lateral PFC know about the conflict between your perception of the word's case and your knowledge of the word's meaning. The lateral PFC then draws upon its understanding of what the task requires—you've been asked to say aloud the case, not read the word itself—and decides that saying "upper" should take precedence. It then sends a signal telling the visual system, which detects case, to get on with its business; this strengthening of the visual system encourages the word recognition system to just shut up. This whole rigmarole is what results in the slight delay and the feeling of effort that you don't get when the word *lower* is actually printed in lowercase letters. In the latter case, the two regions work together happily, the ACC conflict detector is not activated, and the lateral PFC is not called upon to adjudicate between squabbling sets of neurons.

Armed with this information, we can now see how a brain on *wu-wei* might function without our having to invoke Obi-Wan's Force.

We can even get a reasonably precise *picture* of it, thanks to some recent neuroscientific work on *wu-wei*–like states. There are inherent challenges with using brain imaging techniques to study any type of activity, because the technology is typically uncomfortable, invasive, and incredibly bulky. The brain scanning technique that provides the best spatial discrimination—the one that allows us to see most clearly what's firing or not in a person's brain—is fMRI (functional magnetic resonance imaging), which measures blood flow to the neurons. Though imperfect, this functions as a reasonable proxy for neural activation. With fMRI, subjects have to lie perfectly still in a huge, extremely loud metal tube. This is not ideal for studying any kind of real behavior. You could never get a decent fMRI image of Butcher Ding or Woodcarver Qing in action. However, a clever study by Charles Limb and Allen Braun was able to get around some of these limitations to look at professional jazz pianists in action. They designed a special, nonferromagnetic keyboard that could be taken inside the scanner, which is basically a huge magnet, and positioned their subjects so that they could sit with the keyboard resting on their laps. The researchers then had them play in two different conditions. In the first, "Scales," they were required to play, over and over, a one-octave C scale. In the "Jazz Improv" condition, they were asked to remain in the same key but improvise a melody based upon a composition they'd previously been asked to memorize.

The researchers' most striking finding was the pattern of activation when the pianists switched into improvisation mode: widespread deactivation of the lateral PFC and increased activity across relevant sensorimotor systems, the ACC, and the frontal polar portion of a region known as the medial prefrontal cortex (MPFC). The ACC, as we have seen, seems to be responsible for conflict monitoring. It typically works with the lateral PFC to maintain cognitive control. The ACC detects conflict, then calls in the lateral PFC to set things straight. And it is this combination of ACC and lateral PFC activation that appears to produce the subjective feeling of conscious, effortful activity.

What this study suggests is that, in a spontaneous yet high-skill situation such as jazz improv, the ACC remains alert in the background even when the lateral PFC is turned off. This particular neural configuration may correspond subjectively to the kind of relaxed but vigilant mode we enter into when we're fully absorbed in a complex activity. It is the ACC that is still paying attention when Butcher Ding is moving his blade effortlessly through the ox, ready to call for help if he runs into a "difficult spot." In other words, at least some forms of *wu-wei* appear to involve shutting down active conscious awareness and control while maintaining background situational alertness. When your conscious mind lets go, the body can take over.

The early Chinese ideal of *wu-wei* involves just this sort of effortless action, and the body unleashed is impressive to behold. Like Luke Skywalker when he allows himself to be led by the Force, the *wu-wei* exemplars we see in early Chinese texts display an almost supernatural effectiveness as they move through the world. Moreover, all of this is quite plausible from a contemporary perspective. Despite the conscious mind's confidence that it is in charge and knows best what to do, the body actually seems to do just fine on its own. Evolution has off-loaded the vast bulk of our everyday decision making and judgment formation onto our automatic, hot, unconscious system because in most situations it is fast, computationally frugal, and reliable. The social psychologist Timothy Wilson has coined the term the *adaptive unconscious* to refer to the vast repertoire of implicit skills, habits, and perceptions that allow us to navigate the world effectively with little, if any, conscious awareness, and there is now an enormous, and growing, popular literature on the power of unconscious thought. The "thinking without thinking" strategy certainly has its limits, but it is becoming increasingly clear that any behavior we want to be reliable and fast in real-life situations needs to be turned into an embodied habit or tacit skill. As the noted jazz pianist Greg Burk puts it, the best improvisation flows effortlessly from a body freed from conscious thought, where "the choice of notes, the silences, the attacks, and the shaping of each phrase all express a unified and organic whole."

It is worth observing, though, that this effortless, unselfconscious body is not performing in a social vacuum. A jazz pianist plays with, or at the very least for, other people. His or her ability to become completely absorbed in the ebb and flow of the music is also predicated on actually *caring* about it. The same is true of any *wu-wei* activity, whether solitary or in a group. This aspect of *wu-wei*—its dependence on social interaction and shared values—is something that is generally overlooked by modern scientific treatments of spontaneity, and it's a critical omission. The case of a drunken hitchhiker from ancient China can help us see why.

2

Drunk on Heaven

THE SOCIAL AND SPIRITUAL
DIMENSIONS OF *WU-WEI*

AMONG THE STUDENTS WHO TAKE MY COURSE ON EARLY CHI-nese thought, a perennial favorite is a story from the *Zhuangzi* that concerns someone hitching a ride home after a night of heavy drinking. "When a drunken person falls out of a cart," Zhuangzi observes, "although the cart may be going very fast, he won't be killed. His bones and tendons are the same as other people's, and yet he is not injured as they would be." I always advise my undergrads not to try this at home, but I've personally seen the principle in action. Under conditions that remain somewhat hazy to this day, I once witnessed a friend take an incredible tumble down a grassy hill, then a short flight of stone stairs, finally and inexplicably ending up on a concrete sidewalk, on his feet, completely unharmed. The rest of us were so impressed that we tried it, too, all of us landing on our feet like world-class gymnasts and then sauntering back to the dorms.

The explanation for our temporary gymnastic excellence is relatively simple: drunk people are physically (and often morally) looser because, to draw upon the terminology introduced above, their cognitive control regions are partially disengaged—"downregulated," as neuroscientists would put it. They are only dimly aware of what's going on around them. This makes them less inhibited—more willing to do the Macarena, for instance—and also less inclined to consider future consequences. On the negative side, this leads to behavior that is sometimes regretted by the sober, hung-over, coldly conscious self the next day (see Macarena, above). On the positive side, it means that when the inebriated, say, fall off a cart, they don't stiffen up in anticipation of contact with the ground but just roll with the fall and therefore emerge relatively unhurt. As Zhuangzi puts it, his drunken cart-rider "was not aware that he was riding, and is equally unaware that he has fallen out. Life and death, alarm and terror cannot enter his breast, which is why he can bump into things without fear." This, Zhuangzi explains, is because "his spirit is intact."

Why does Zhuangzi celebrate the partial paralysis of our cognitive control regions? Because drunkenness induces something very much like a crude form of *wu-wei*: a short-term suspension of active self-monitoring. At the same time, it's clear that this metaphor of the drunken man is just that—a metaphor—and that the substance Zhuangzi *really* wants us to get drunk on is Heaven. "If a person can keep himself intact like this by means of wine," he concludes, "how much more so can he stay intact by means of Heaven! The sage hides in Heaven, and therefore nothing can harm him." The cart rider is meant to show how being "drunk on Heaven" provides a degree of independence from the ordinary world. Things that would hurt most people do not harm the Heaven-drunk sage.

We see here a tight connection between *wu-wei* and a specific religious worldview, and this is significant. In the previous chapter, we've gone some way toward providing a naturalistic account of at least certain aspects of *wu-wei*—that is, a scientific explanation of how behavior can be effortless, unselfconscious, and yet perfectly efficacious. At the same time, we miss some crucial features of *wu-wei* by too abruptly pulling it out of Warring States China and plopping it into the modern world. For the early Chinese, being in *wu-wei* is not just about how one feels internally, or to what extent one's conscious brain is in charge. It is, at the end of the day, about being

properly situated in the cosmos. And this too has important implications for contemporary life.

Everyone—including the early Chinese—loves the Butcher Ding
story. Everyone—including the early Chinese—also thinks that
being physically skillful and effective is a great thing. Ultimately,
though, thinkers like Zhuangzi don't really care about how you
cut up oxen or carve bell stands. What they care about is how you
relate to others. This is why Butcher Ding says, at the end of the
story, that what he cares about is the Way (the *Dao*), not skill per se.
After seeing his performance, Lord Wenhui doesn't announce that
he'd really like to give up the feudal lordship business and become
a butcher, but rather that he's learned from the performance the
secret to living his life. This is because the story is really about *social
effectiveness*: the ability to move through the human world with the
same ease as Butcher Ding's blade through the ox.

This social dimension of *wu-wei* becomes crystal clear in texts
such as the *Analects*, where we read that Confucius always appears
in the proper attire, knows the right way to enter a room, utters
the most apropos words, and deals with others in the most tact-
ful manner, and yet is never stiff or formal. His behavior seems to
flow spontaneously from his very nature, which completely disarms
philosophical opponents, tames ornery rulers, and shames lazy disciples into pursuing their studies with renewed effort. You can practically hear the *swish* and *swoosh* of Ding's blade as Confucius glides
through an official reception with consummate grace or deftly han-
dles a rude interlocutor without missing a beat.

This kind of social efficacy, in turn, depends crucially upon the
mysterious power of *de*, the charismatic tractor beam that emanates
from a person in *wu-wei*, drawing other people to him or her and inspiring trust. In the *Analects* we read that "one who rules through
the power of *de* is like the Pole Star: it simply remains in its place
and receives the homage of the myriad lesser stars." In early Chinese astronomy the Pole Star was thought to be the fixed center

of the nighttime sky, with all of the other heavenly bodies moving around it in concentric circles. The truly virtuous Confucian ruler can simply take his place in the palace and the gravitational force of his *de* will draw everyone into his or her proper place around him. In the *Laozi* we read of the hidden Daoist adept, with an empty mind and a child's heart, who is able to walk among fierce wild animals or through walls of fire without being harmed. The Laozian ruler is also able to bring the human world into order through his *de*, although unlike the Confucian ruler—high up and bright in the heavens like the Pole Star—the Laozian sage is invisible, dwelling in the dark valleys and pulling everyone in the world into order as gravity pulls water downhill. The Zhuangzian sage similarly possesses a powerful *de*, which has a therapeutic, relaxing effect on others and allows him to move unharmed through the physical and social worlds.

For the early Chinese, there was a fundamental link between this crucial power of *de* and Heaven, the drink of choice for our Zhuangzian sage. The concept of Heaven (*tian*, 天) is an ancient one, appearing in bronze vessel inscriptions from the second millennium B.C.E. It refers to the high god worshipped by the early Zhou kings, conceptualized as a powerful person dwelling somewhere up in the sky—*tian* refers both to this being and to the physical sky. This is why "Heaven" is the standard English translation, and probably the least inaccurate one we can find. It's important to realize, however, that we're not talking about a *place*, as in the Christian conception of heaven, but rather a godlike being who can send down orders, control the weather, determine success in battle, and reward and protect its followers. From Zhou times on, Heaven is also seen as the source of value or goodness: what Heaven wants is, by definition, good. The same inherent goodness characterizes the "Way" (*Dao*, 道), which literally refers to a path or road—a physical *way*. By extension, it can also mean the *way* to do something and, in this context, the *right* way. For the early Chinese, it had cosmic significance: the Way is the proper means of being a perfected human being, or faithfully doing the will of Heaven. The Way is Heaven's Way, the grounding of all goodness or value in the world.

So *wu-wei* and *de* are fundamentally linked to Heaven. *Wu-wei* works because to be *wu-wei* means that you are following the Heavenly Way, and anyone following the Way gains the power of *de*. This connection has important implications for understanding the kind of spontaneity that the early Chinese valued so highly. Spontaneity in the West is typically associated with individuality—people just doing whatever they want. *Wu-wei*, on the other hand, means becoming part of something larger: the cosmic order represented by the Way. Sages from Confucius at age seventy to the Daoists describe *wu-wei* as a state of "fitting" with the universe. Similarly, *de* is powerful because Heaven has made humans, animals, and even the natural world in such a way that they respond instantly and unquestioningly to virtue. The *de*-bearing sage can attract people, calm wild animals, and ensure good harvests and clement weather. By rewarding the *wu-wei* ruler with this power, Heaven ensures that its will is done. *De* is like a halo that surrounds someone in *wu-wei* and signals to everyone around: "Heaven likes me! You should too! I'm okay."

Grasping the fundamentally religious nature of *wu-wei* is vital, and not just for reasons of historical accuracy. For one thing, the example of mastering a physical skill (like ox butchery), while a helpful analogy, becomes misleading if it's disengaged from its original cultural and religious context. The problem is that we can imagine someone being a skilled butcher, pianist, or tennis player and yet still an atrocious human being. (I'm sure you know a person or two like this.) What *wu-wei* represents, on the other hand, is the state of being a perfected part of a greater whole that is also embraced by others. It is this holistic, social, and religious quality of *wu-wei* that makes it unique.

The best way to get a handle on this point is to consider what is probably the best-known contemporary account of spontaneity as an ideal, the concept of "flow" developed by the psychologist Mihaly Csikszentmihalyi. If we consider the major features of flow as he outlines them, many similarities pop out: deep but effortless concentration, responsiveness to the environment, a high degree of effectiveness, profound enjoyment, the loss of a sense of self, and an altered sense of time. Over the past couple of decades, Csikszentmihalyi and his colleagues have demonstrated that flow experiences appear to be universal cross-culturally and to be described in similar ways by people involved in a broad range of activities. In Csikszentmihalyi's view, the key feature that unites all of these experiences— the defining condition for flow—is a precise calibration of challenge and skill. Flow happens when we hit the "sweet spot" between too easy and too hard. Since skills improve over time, this means that flow requires exposure to constantly "spiraling complexity" that "forces people to stretch themselves, to always take on another challenge to improve on their abilities." It is this emphasis on *challenge* and *complexity* that best allows us to see the distinction between flow—at least as defined by Csikszentmihalyi—and *wu-wei*. It also helps us to see how Western individualism can obscure certain important aspects of spontaneity that are, in contrast, highlighted by the early Chinese.

Consider one of the examples that jumped out at me when I first read *Flow* as a college student, that of "E.," an anonymous but reportedly well-known and powerful European woman:

A scholar of international reputation, she has at the same time built up a thriving business that employs hundreds of people and has been on the cutting edge of its field for a generation. E. travels constantly to political, business, and professional meetings, moving among her several residences around the world. If there is a concert in the town where she is staying, E. will probably be in the audience; at the first free moment she will be at the museum or library. And while she is in a meeting, her chauffeur, instead of just standing around and waiting, will be expected to visit the local art gallery or museum; for on the way home, his employer will want to discuss what he thought of its paintings.

The example of E. certainly fits the profile of individual challenge and constantly spiraling complexity. From what we can glean, she is an impressive individual. Her life story is inspiring, and she herself "radiates a pure glow of energy." I have to admit, though, that when I first read about E. her lifestyle struck me as rather exhausting, with all of this constant traveling, activity, and relentless self-improvement. (I also remember thinking that, were I her chauffeur, every now and then I might just want to chill out and have a smoke instead of being force-marched through a museum and then getting quizzed about it.)

Flow also contains other stories that I originally found much more compelling than that of E. One good example is the story of Serafina Vinon, a seventy-six-year-old inhabitant of a small mountain village in the Italian Alps. Serafina is described as living a lifestyle that remains more or less unchanged from that of her ancestors for the past few hundred years: she rises at five in the morning to milk her cows, she cooks, she cleans, she takes the herd up to higher pastures or tends to her orchards, she cards wool. When asked to describe what gives her the greatest pleasure in life, she reports that it is precisely these everyday activities that put her into a state where she is absorbed, unselfconscious, and at ease. "To be outdoors, to talk with people, to be with my animals . . . I talk to everybody— plants, birds, flowers, and animals," she explains. "Everything in nature keeps you company; you see nature progress every day. You feel clean and happy. . . . Even when you have to work a lot it is very beautiful."

This story has little to do with complexity and challenge, still less with constantly *spiraling* complexity. Rather, the main sense one gets is of peaceful, relaxed absorption into something larger than the self: the natural beauty of the Italian Alps, an inherited tradition that provides routine and structure, a sense of companionship with animals and birds. These bigger wholes into which Serafina feels absorbed—Nature with a capital "N," or her cultural tradition—are also seen by her as important sources of value or goodness. Indeed, it is the fact that she deeply *cares* about her beautiful surroundings and her way of life, and derives meaning from them, that allows her to become absorbed in the first place. When I first read this account, as a beginning student of Chinese thought, I recall feeling that Serafina's experience not only looked a lot more like early Chinese conceptions of *wu-wei* than the endless travel and self-improvement of E. but also matched my own experience better. This is a conviction that has only strengthened over time.

To be sure, when I'm lucky I am able to enter *wu-wei* in complex, challenging situations linked to my job—for instance, while writing this book. When I'm in the "writing zone," as I think of it, I experience all of the classic flow hallmarks. Entire paragraphs seem to emerge spontaneously from some mysterious region of my brain, I am completely absorbed in what I'm doing, I lose all sense of time, I forget to eat, and I emerge feeling relaxed and pleased with my work. No one could deny that complexity and challenge play a role here. In the final analysis, though, they are incidental. The complexity and challenge of writing a book can induce *wu-wei* only if they are encountered in the service of something bigger, such as an idea that I really care about and want to share with others.

It is this focus on *caring*—on getting beyond the self—that, in turn, allows us to connect *wu-wei* states characterized by high complexity and challenge to their infinitely more common relatives: very routine, thoroughly familiar, low-complexity activities that allow us to be fully absorbed in something that we love and value and that we see as being larger than our individual selves. My most common *wu-wei* experiences, like Serafina's, have always tended to involve activities that put me in contact with the natural world. Hiking the same trail I've hiked a hundred times through Point Reyes National Seashore on the California coast reliably puts me in *wu-wei*, as does weeding my vegetable garden, pruning and tending to my fruit trees, or simply sitting on an ocean beach watching the waves crash. Except for the psychological profile of the state they produce, these activities do not look anything at all like E.'s frenetic life. Indeed, if one looks past Csikszentmihalyi's own emphasis on complexity and challenge, the survey data that he and his colleagues have gathered suggest that most flow experiences actually occur in social situations of relatively low complexity, like conversing with friends, sharing a meal with family, or playing with small children.

Why, then, did flow researchers end up focusing on complexity and challenge as the defining features of the experience? Because these researchers are, for the most part, Western individualists. In our culture, activities like running ultramarathons or exploring new art museums tend to be solitary and aimed at self-improvement. We focus more on the intersection of our individual skills with the demands of the task at hand and overlook the fact that the resulting challenge engages us only against a broader background of things we care about. Flow researchers also latched on to complexity and challenge because they were keen to distinguish flow states from other states that share some of the same features—loss of a sense of self, an altered experience of time, relaxation—but that we wouldn't want to dignify with the label of "flow." Watching a mindless TV show would be a good example, but we could include any simple, passive activity that fully occupies our attention while we are immersed in it, yet leaves us feeling empty or tired rather than fulfilled and energized: gambling, reading a trashy magazine, aimlessly surfing the Internet, or exchanging desultory gossip with acquaintances we don't even really like. When faced with the question of how to distinguish between flow and activities such as this, complexity and challenge jump out as the salient features when considered from a purely individualist perspective.

Our consideration of the early Chinese religious worldview, however, gives us an alternate, and ultimately more satisfying, way to make this distinction. It is the connection with a larger, valued whole that allows *wu-wei* or true flow experiences to leave us feeling "clean and happy," as Serafina puts it, rather than dirty and worn out. One can think of this larger whole as a *framework of values*—that is, a structure within which we situate ourselves or our actions and that allows us to classify some things as *good* and others as *bad* and to behave accordingly. Many in the West have attempted to ground such values in objective facts or rational calculation, but they are, by their very nature, beyond the purview of science. Science can tell us what *is*, not what *should* be: it traffics in facts, not values. This means that we go beyond the facts anytime we make value judgments. We feel that slavery is *wrong* because there is something special about human beings that makes us different from cattle, although from a biological perspective there is nothing that qualitatively distinguishes *Homo sapiens* from *Bos taurus*, the domestic cow. Our value judgments are ultimately grounded in these kinds of unjustifiable, but nonetheless deeply held, convictions. Moreover, human beings are built in such a way that we cannot do without such convictions. Imagine trying to move through life without any sense of good or bad, or without the motivational drive that such commitments give us.

Understanding the essential role that these frameworks play in human life is, in turn, the key to understanding *wu-wei*. I would suggest that the distinguishing feature of *wu-wei* is the absorption of the self into something greater. That is, whether we emerge from a state of effortlessness and unselfconsciousness feeling energized or enervated probably depends, at least in part, on our values: How does the activity in which we just engaged reflect our larger sense of who we are and what we hold dear? For example, if you value a particular set of friends, you're likely to come out of an evening of drinking and chatting with them feeling good about how you've spent your time. The mere anticipation of value may also be a facilitator of *wu-wei* experiences: the fact that Point Reyes occupies a special place in my own personal set of values makes me much more prone to experience *wu-wei* when hiking there. Indeed, the very act of preparing to go there puts me into *wu-wei*, and even the drive up takes on special qualities.

For most of recorded human history, value frameworks were provided by organized religion and were shared more or less universally across large populations. For the early Chinese thinkers, for instance, this framework was provided by faith in Heaven and its Way. For traditionally religious people today, this "valued whole" looks very much like the early Chinese Way: a coherent, clearly defined metaphysical structure (God's plan, the working out of karma), backed by a priesthood and a set of important texts, that gives meaning to activities that contribute to a greater good. Such frameworks typically include a widely shared sense of which places, objects, and activities are "sacred," or endowed with supercharged meaning. Another relevant feature of traditional religions is that they tend to really script out your day. There are loads of rituals to be performed at specified intervals, concrete guidelines about what to eat and wear and touch, and lots of group activities. This makes interacting with the sacred, typically in the company of like-minded individuals, a very reliable and effective means of getting into a state of *wu-wei*. Indeed, this is probably how *wu-wei* has been experienced by most people for most of human history.

We are now living in an age, however, where traditional religious values and commitments are being called into question. Many people today have rejected all traditional religious frameworks, characterizing themselves as secular or "spiritual but not religious." This is not to say that they have figured out how to live entirely without value frameworks. If we take a step back, we will observe that even the most ardently atheistic secular humanist is still committed to *some* very abstract, metaphysical framework, revolving around such values as a respect for human dignity and freedom, individualism and equal rights, and rationality as the preferred means for guiding public policy. Looked at this way, secular humanism functions very much like a traditional religion. It gives its followers a set of sacred values that allows them to make distinctions between right and wrong, good and bad, as well as the motivation to punish or shun those who do not live up to those values.

In another sense, however, contemporary secularism has new and distinct features, the source of both its strengths and its weaknesses. Its commitment to rationality and evidence, for instance, means that it's unusually open to modification around the edges, although—as with any value system worth its salt—the core values like human rights or freedom are in principle non-negotiable. The flip side to this openness is a somewhat disorienting minimalism: liberalism is about as stripped down as a value system can be and still function. Most of its injunctions are *negative*. Do not violate human rights, do not restrict people's freedom of expression, do not allow the strong to oppress the weak. As long as you are careful to steer clear of committing genocide or being oppressively prejudiced, however, secular humanism then doesn't have a lot to say about what you *should* be doing. Besides vaguely sacred communal rituals such as listening to NPR, reading the *New York Times*, or buying locally sourced organic vegetables, secular humanists are not given much guidance on how to actually live their lives. And this vacuum has to be filled by something—avoiding human rights abuses still leaves a lot of hours in the day.

This is why we tend to align ourselves with certain social tribes—Suburban Soccer Mom, Urban Hipster, Tortured Artist—that can live comfortably under the very large, but rather empty, umbrella of secular humanism. These groups or stereotyped roles give us precisely the kind of detailed scripts that secular humanism eschews, specific guidance regarding dress, food, and other lifestyle details normally provided by traditional religions. Tortured artists dress strictly in black and are encouraged to get small, esoteric tattoos, to smoke, and to read Camus. Minivans are the sacred chariot of the Soccer Mom but anathema to the Urban Hipster, unless driven ironically. (For Urban Hipsters, irony functions like a magic shield, allowing them to emerge unscathed, or even invigorated, by contact with objects or scenarios—meatloaf, diners, kitschy movies, seventies-style clothing—that would otherwise be considered *trafe* or unclean.)

In addition to the everyday guidance provided by these tribes, which can be mixed and matched, hybridized, whatever floats your boat, secular people can often be found embracing a variety of more specific value commitments—such as environmentalism, economic reform, or hedonism—that can all be accommodated by the bigger framework of secular liberalism. Hedonism, to focus on my personal favorite, centers on the belief that value lies in the maximization of pleasure, broadly understood. The word *hedonism* itself refers originally to a Greek school of thought, but despite its modern connotations the early Greek hedonists were actually not all that fun. You certainly wouldn't want to party with them. In the early Greek hedonist view, most of the things—like sex, food, or wine—that are typically viewed as pleasurable by the "vulgar" (that is, you and me) are in fact causes of suffering in the long run. This is because they are by their very nature ephemeral and therefore never genuinely satisfying. The only way to truly maximize pleasure, in the Greek hedonists' view, was to stick to eternal, imperishable pleasures, like philosophical reflection, while keeping one's involvement in the physical world to a bare minimum.

As a hedonist, I'd probably peg myself somewhere in between the modern and ancient Greek versions, but closer to the modern. I get the point about the enduring pleasures of the mind but have a more optimistic view about food and sex and wine than the ancients did. Sheltering under my broader, but pretty abstract, secular liberalism—with its commitments to human rights, individual freedom, evidence-based public policy—my smaller personal faith, if you could even call it that, is anchored in a set of personal relationships, an enjoyment of intellectual inquiry, a profound love of the ocean, an appreciation of good food and wine, and an oddly firm sense that human beings are meant to live only in Mediterranean climates with access to fresh citrus and proper olive oil. In fact, I used to have a litmus test for places I'd be willing to move to: if you couldn't grow lemon trees there, it was off the list.

That said, I've now lived in Vancouver, Canada, for over six years, and have done so quite happily, despite the tragic death of my small potted lemon tree halfway through my first Pacific Northwest winter. This just goes to show how coherent and consistent my personal framework is. If you press me on it, I'll have to admit that I have no coherent—let alone empirically defensible—story about, say, what makes Point Reyes special. But it *is* special for me, and this halo of specialness extends to the other aspects of my life that I value: my family, my friends, certain landscapes and tastes and colors. A very common predicament for people like me is to experience *wu-wei* in the context of small-scale, fleeting moments where I feel at home and connected to something. If we're lucky and have managed to set up our lives well, we spend most of our time moving between these moments and would describe ourselves as "happy," although if there's any overarching belief structure it's typically pretty minimal.

As silly as these fragmentary collections of personal frameworks may seem to a traditionally religious person, they are all we secular folks have in terms of concrete commitments, and they *do* seem capable of inducing *wu-wei*. By dint of great effort, I somehow manage to tenuously maintain a Mediterranean-style garden above the forty-ninth parallel, and tending to that garden—or even gazing at it through my office window as it bravely endures the inhuman rain and darkness of a Vancouver winter—will often put me into a state of *wu-wei*. The same is true of spending time with my wife and daughter, having a drink with friends and colleagues after work, deciphering a difficult ancient Chinese text, figuring out what makes human beings build pyramids or stick metal skewers through their cheeks, or hiking the beautiful, steep path between my campus office and the rugged beach fronting the Strait of Georgia. Interestingly, if you were to keep track of the activities that induce *wu-wei* in me or anyone else, you'd be able to piece together a rough outline of what sorts of things a person values or doesn't. You'd also be able to tell *whom* they value, and whom they don't, and this is perhaps of even greater importance. Crucially, *wu-wei* can occur in group activities only when we genuinely value the social relationships involved. We can effortlessly engage with others only when we care about, and feel relaxed with, the people we are with.

It's now easy to see why *wu-wei* is about more than isolated individuals incrementally improving their personal bests in the Ironman Triathlon or mastering a new level of Tetris. *Wu-wei* involves giving yourself up to something that, because it is bigger than you, can be shared by others. For those of us who no longer embrace the early Chinese faith in the Way and Heaven, the precise nature of this "larger whole"—the framework of values that gives shape and meaning to the *wu-wei* experience—is going to vary from tribe to tribe, even person to person or moment to moment. By its very nature, though, this framework needs to be something larger than the self. An essential fact about *wu-wei* is that it's not just about the experience unfolding within the mind of an isolated individual but also about social connections *between* people. This has some important social and psychological consequences that we will explore later.

So, we now have a sense of how *wu-wei*'s fundamentally spiritual and social nature reveals aspects of spontaneity that modern psychological approaches have overlooked. We still, however, have not addressed the bigger question posed at the beginning of the book. Being in *wu-wei* is great. Having a bit of *de* to smooth our way through the social world is obviously desirable. How, though, do we get these things if we don't already have them? How do we try not to try?

One of the great strengths of the early Chinese thinkers is that they did not merely describe states of effortless perfection but also focused on how to create these states, developing a variety of cultural practices, mental techniques, and physical exercises meant to nudge us into the right sort of spontaneity. As we'll see, they tended to oscillate between *trying* strategies (work really hard, and eventually you'll acquire *wu-wei*) and *not trying* strategies (just stop trying, and *wu-wei* will be there). We'll begin with Confucius and his follower Xunzi, who developed the first and greatest of the *trying* strategies, and the one against which all subsequent strategies were formulated.

3

Try Hard Not to Try

CARVING AND POLISHING THE SELF

THERE ARE MANY FAMOUS PASSAGES FROM THE *ANALECTS*, THE collected teachings of Confucius, sayings that everyone can recite from memory and that stand out for their pithiness and profundity. One of my favorites, however, is a rather obscure passage you'll never see on a T-shirt, a deceptively straightforward account of how Confucius behaved when receiving a blind master musician as a guest. The post of musician was traditionally filled by blind persons in ancient China, both to give them a trade in which they could excel and because their sense of hearing was considered more acute than that of the sighted. This particular master musician has presumably been brought to Confucius's residence by an assistant, who then leaves him in Confucius's care. In ancient Chinese society, musicians—even master musicians—were not like contemporary rock stars or opera divas: they were relatively low-ranking functionaries, expected to remain in the background. Despite his relatively humble social rank, however, this master musician is immediately greeted by Confucius and taken under his personal care, with remarkable solicitousness and sensitivity:

> When they came to the steps, Confucius said, "Here are the steps."
> When they reached his seat, Confucius said, "Here is your seat."

After everyone was seated, the Master informed him as to who was present, saying, "So-and-so is seated here, and So-and-so is seated over there."

When the master musician left, a disciple asked, "Is this the way to converse with a master musician?"

Confucius replied, "Yes, this is indeed the way to assist a master musician."

Despite the disciple's question, this passage is intended, not as an instruction manual for dealing with a particular rank of guest, but rather as a chance to show an *wu-wei* Confucian gentleman in action. Note the economy of expression of Confucius, who puts aside the normal ritual behavior of a host—typically rather standoffish and restrained, especially when dealing with a social inferior—in order to deftly and respectfully guide his blind guest, without being overly fussy or condescending.

This passage, however, is also potentially misleading, because it seems to describe the behavior of a person, Confucius, who just happens to be a really nice guy. Our tendency when we come across someone who is skillfully and effortlessly graceful is to think that they must be *born* that way. The *Analects* makes it very clear, however, that this is not at all how we are to understand Confucius. Confucius repeatedly describes himself as not particularly gifted but merely someone who "loves the ancients" and has devoted his life to learning the Way handed down by the Zhou kings. Confucius despaired about the world that he lived in, a fragmented China about to descend into the three centuries of chaos that constitute the Warring States period. He looked back to the early Zhou dynasty (roughly 1000–700 B.C.E.) as a lost Golden Age, when the known world was unified, peaceful, and in harmony with Heaven. The key to regaining this harmony, in his view, was to relearn ancient cultural practices—the Way of the Zhou kings, revealed to them directly by Heaven—that had been forgotten or neglected.

Confucius was convinced that *no one* was born graceful yet

correct, and to the end of his life remained fond of rebuking anyone who thought that untutored casualness was desirable or even socially acceptable. This point is well illustrated by an *Analects* passage that describes Confucius's reaction upon entering a room and being welcomed by a young man named Yuan Rang. We don't know much about this Yuan Rang, but he seems to have been an early Chinese hippie, possibly a devotee of the Laozi we'll meet in the next chapter. I've always imagined him with dreadlocks and loose robes, smelling vaguely of patchouli. Confucius enters the room and finds this guy "sitting casually, with his legs sprawled out, waiting for Confucius." The proper attitude for receiving a guest in a situation like this would have been the formal kneeling posture still practiced by the Japanese: kneeling on a mat while sitting back on the ankles. Yuan Rang was apparently lounging with his legs stretched out or loosely folded in front of him, a position that is certainly more comfortable but rather too casual for receiving an elder such as Confucius. The Master's response is priceless: "Upon seeing him, the Master remarked, 'A young man devoid of humility and respect for his elders will grow into an adult who contributes nothing to his community. Growing older and older without the dignity to simply die, he becomes nothing more than a useless burden.' He then whacked him on the shin with his staff." I can't tell you how many times I've wished for both a staff and permission to use it when confronted with slovenly students openly chatting with their friends or noisily scarfing down food during class. An emphatic whack on the shins would do them a lot of good. University lawyers have counseled me against this, however, so I have to be content with simply citing this passage and hoping the point somehow sinks in: learn some manners.

Despite popular images of wise old Daoist recluses relaxing in the countryside, playing desultory games of chess and drinking wine, by far the most dominant strategy for attaining *wu-wei* in traditional China was the one first laid out in the *Analects* of Confucius and then elaborated in much more detail by his follower Xunzi at the end of the Warring States period. For them, *wu-wei* could be obtained only through a lifelong program of *trying*, strenuously, to distance oneself from nature and move instead toward a thoroughly cultivated kind of perfection. This is because both of these thinkers were convinced that our inborn tendencies, if indulged, would lead to ugly consequences. In their minds, the only way to achieve a ful-filling life and social harmony was to reshape our nature in accor-dance with cultural ideals inherited from the past. Their aim was still *wu-wei*, but this was understood as a kind of *artificial* spontane-ity, a cultural and educational achievement rather than the result of simply going with the flow.

Another way to put this is that, despite their endorsement of spontaneity as an ultimate end goal, Confucius and Xunzi empha-sized *cold* cognition, on both an individual and a cultural level. For individuals, they stressed the importance of exerting willpower, con-sciously reflecting on one's behavior, and repressing hot cognition when appropriate—which, in the early stages of training, is almost always. The cultural forms that they celebrated can be seen as a crys-tallization of acts of cold cognition performed in the past, a reservoir of wisdom produced by careful, conscious reasoning. By master-ing and internalizing the cultural ways inherited from the ancient kings, otherwise misshapen human beings could transform them-selves into something beautiful. Although their reliance on cold cog-nition might sound similar to Western rationalist philosophy, it is the Confucian emphasis on personal *transformation* that marks the crucial difference, as we will see.

No single person could hope to reproduce this inherited wisdom on his own. As Confucius puts it, "I once engaged in thought for an entire day without eating and an entire night without sleeping, but it did no good. It would have been better for me to have spent that time in learning." Thinking on one's own might be compared to ran-domly banging on a piano: a million monkeys given a million years might produce something, but it's better to start with Mozart. We find a very similar theme in the *Xunzi*:

I once stood on my tiptoes to look into the distance, but this is not as good as the broad view obtained from climbing a hill. Climbing a hill and waving your arms does not make your arms any longer, but they can be seen from farther away; shouting downwind does not make your voice any louder, but it can be heard more clearly; someone who borrows a carriage and horses does not improve the power of his feet, but he can travel a thousand miles; someone who borrows a boat and paddle does not thereby become able to swim, but he can cross great rivers. The gentleman by birth is not different from other people—he is simply good at borrowing external things.

These "external things" are the various cultural practices passed down by the Zhou kings—the fruits of cold cognition inherited from the past. Xunzi wanted to emphasize that the kind of naturalness that Confucius valued doesn't just fall off the trees but is a hard-won achievement. Put another way, he wanted to emphasize that hot cognition needs to be restrained and then extensively reshaped by cold processes before it can be trusted. The evolved structure of our embodied minds suggests that Xunzi was on to something here.

HOT IS NOT ENOUGH: WHY WE HAVE CONSCIOUSNESS AND CULTURE

As we've seen, recent work in cognitive science has tended to highlight the power of hot cognition. It is fast, frugal, effective, and clearly in charge of most of what we do in the world. The picture that seems to emerge from this literature is of consciousness as cheerleader, jumping around, making a big to-do about events on the field that it played no causal role in bringing about. This raises an obvious question: given the cost and slowness of our conscious mind, and the speed and efficacy of our unconscious mind, why bother with the whole consciousness thing at all?

Recent scientific work offers some answers to this question and, in doing so, sheds a great deal of light on the advantages of the Confucian strategy. Cognitive scientists have demonstrated that consciousness pays for itself by allowing us to pull off a variety of impressive tricks. For one thing, it allows for flexibility when hot, fast, automatic skills hit a roadblock. As we've seen, when the unconscious mind is stymied, it sends out an SOS (via the ACC) to the conscious mind, asking it to spring into action, figure out what's wrong, and try to fix it. Consciousness is also called in as referee when unconscious drives find themselves in conflict (desire for a sweet conflicting with a desire to go to sleep) and when overriding our unconscious desires would be in our long-term interests (desire for a sweet incompatible with our decision to lose weight).

Besides helping us stick to our diets, it's also clear that consciousness is vital when it comes to managing our social lives. Our unconscious minds are very good at quickly detecting agency, identifying threats in our environment, and reading emotions in faces. Only conscious processes, however, seem capable of complex *modeling* of other minds. Consciousness creates a virtual representation of the internal thoughts of others so that we can figure out how to interact with them: he is disappointed I didn't say hello, she is wondering why I'm so late. The virtual world of consciousness is also where we get to practice things without actually having to *do* them. You've been admiring that woman for a while, she seems to like you, you see her every morning, what are you going to say when you sidle up to her in the line at the coffee shop tomorrow? Practice it in your head now. Offline, imaginary practice is equally important when it comes to physical skills. Anyone learning to ride a bike or hit a golf ball mentally practices what she's learned, rehearsing the moves over and over, and this helps to make the artificial, novel movements involved more fluent and effective.

Consciousness is also the seat of language, an essential tool for social animals like us. While language has its limits—trying to explain what a wine tastes like, or how an emotion feels, is notoriously difficult—it is extremely effective in converting personal experience into a medium that can be shared with others. You can, at least haltingly, explain to me why you prefer this Chardonnay to that one, or why you're so angry with me. Language is also crucial for the narration that creates our continuous sense of self. The ego has an annoying tendency to take credit for the accomplishments of hot cognition, but without the sense of narrative unity that it provides we'd be completely at sea, with no idea of who we are, where we've been, or where we're headed. Powerful hallucinogens can, by temporarily paralyzing the areas of our brain responsible for orienting us in space and in language, give us a sense of what life without the language-based ego would be like. Full of color and magic, perhaps. Being high on mushrooms makes it rather hard, though, to get anything done, or to communicate an idea clearly. This is why attempts by people tripping on hallucinogens to record their wondrous and profound experiences—in written words, art, or multimedia— typically end up appearing trite and incoherent. If you doubt this, take a look at some of my journal entries from the early 1990s in which I definitively prove, with extensive diagrams and flow charts, that Truth is the color blue. Drug-inspired writing, when it works (Jack Kerouac and Aldous Huxley come to mind), does so only because it's been extensively reorganized and translated into something intelligible by the plodding, sober, conscious mind.

Language is also clearly crucial for a special kind of abstract thinking that requires what the philosopher Daniel Dennett calls "scaffolding"—chains of reasoning so complex that we need external placeholders to keep us from forgetting where we are. This scaffolding includes such useful cultural inventions as, say, calculus, statistics, or controlled double-blind experiments, which allow us to uncover patterns in the world that are completely invisible to our hot cognition. The creation of modern science is essentially a story of how, over a long period of time, humans have cobbled together novel methods of thinking and communicating that allow us to reach conclusions that completely contradict our intuitions—

born of hot cognition—but give us a more accurate picture of how the world works. The earth moves around the sun. Colds are caused by tiny, tiny organisms, not cool breezes or uncovered feet (although try telling this to my Italian mother-in-law).

The sum total of these scaffolding devices can be referred to as "culture": a body of information passed down from generation to generation in a process that, in some important respects, resembles genetic evolution. One of the key features of this process is that it can generate solutions that are, in principle, beyond the reach of any individual or single generation. For instance, minor but helpful variations in tool design can be "selected for" in the sense that some designs survive and some don't, even though no actual person ever consciously decides to prefer one version over another. Similarly, cultural evolution can zero in on important long-term problems that are, by their very nature, invisible within a single human lifetime—say, the effect of vitamin deficiencies in local foods—and "design" solutions for them.

To take one example, Fijians have a set of food taboos that forbid pregnant and breast-feeding women from eating certain species of reef fishes. This might appear, within the space of a particular generation, arbitrary and silly. Why shouldn't a pregnant woman have a bit of tasty barracuda or rock cod if she wants it? Researchers have demonstrated, however, that these taboos selectively target species that present a high danger of ciguatera poisoning, caused by the accumulation of toxins in certain species at the top of local food chains. Taboos against eating such species significantly reduce the risk of women experiencing ciguatera poisoning at a particularly vulnerable period of their lives, and when their developing or newborn child is also at risk of poisoning. Importantly, the causal link between consuming barracuda, experiencing ciguatera, and giving birth to a less healthy child has not necessarily been clearly recognized or articulated by any single individual within the community. Avoiding tabooed fish is just what people have come to do in this culture. People have a bias toward imitating successful individuals. Combine this with the fact that women with large families and more successful offspring—who also tend to be those who avoid particular reef fish at crucial points in their pregnancy and nursing—are particularly esteemed in Fiji, and the result is an adaptive set of food taboos that has been unconsciously assembled by cultural evolutionary forces over the course of multiple generations.

As in Fiji, the "collective mind" of any cultural group, accumulated over time, is typically smarter than any individual human mind. This is why cultural learning is so important, and also why such techniques as crowdsourcing are so effective. Xunzi compares the Confucian Way inherited by his generation to markers used to indicate a ford over an otherwise deep and swift river. People with experience have, through careful trial and error, figured out the best place to cross the river and have left markers to help us find it. We could ignore them and just wing it, but that would be counterproductive and even dangerous. In other words, if a respected member of the local community tells you to boil this root vegetable for two hours, then strain it, and then pound it with a stick blessed by a priest until you've sung this sacred song twenty times, you should probably just shut up and do it, exactly the way you are told.

Our ability to rely on what we might call *personal* cold cognition (individuals using their cognitive control regions) and *borrowed* cold cognition (the fruit of many of these conscious acts embedded in our cultural traditions) means that, unlike almost every other species on the planet, we are not prisoners of our unconscious. We can process otherwise dangerous foods that no other animal can eat. We can pass down information about where to find game at different times of the year, how to build a seaworthy kayak, and how to best manage disputes between hunting partners. This has allowed humans to inhabit practically every ecosystem on earth. But living in the kind of complex social groups that cultures tend to give rise to requires more than cold cognition, a crucial point that the Confucians understood quite well.

COLD CAN'T GO IT ALONE: BUILDING
COLD INTO HOT

The idea that "the body"—the panoply of desires and instincts and unconscious habits rooted in hot cognition—is a barrier to perfection is found in all of the world's major religions, as is the conviction that the only way to subdue the body is to use "the mind." We certainly see this theme at work in Confucian thought. Xunzi, for instance, is fond of the metaphor of the mind as ruler. (The Chinese word we translate as "mind" actually refers to a concrete organ in the body, the heart, and so we also sometimes render it as the "heart-mind.") For the early Chinese, the heart-mind was the seat of certain emotions, but its most important function was as the seat of cold cognition and conscious willpower. While the other organs can function only in a hot manner, the heart-mind is capable of stopping, reflecting, weighing pros and cons, and then making a conscious, considered decision. Because of these unique powers, it was viewed as the natural ruler of the self.

This metaphor of mind as master of an otherwise unruly body is a common one. Plato's mind, for instance, is a charioteer, trying to control a team of wild horses that represent hot cognition in various guises. In China, a more common metaphor has been water control: an irrigation manager trying to channel water to where it is needed, or divert it in order to avoid flooding. This makes a great deal of sense for a culture completely at the mercy of the powerful, and erratic, Yellow River, which was a source of water and fertility for the entire Yellow River Valley but was prone to jumping its banks and causing widespread devastation. In all of these folk and religious models, "we" (the charioteer, water manager) need to exert force to control the hot power of nature (animals, water). Although this dichotomy between a rational person and irrational force captures our internal experience of self-control, it's important to keep in mind that it isn't really accurate from a scientific perspective. Self-control is *all* about the embodied mind, in that certain regions of the brain strengthen some neural pathways at the expense of others. We identify with the cognitive control regions, and personify them as rational agents, only because they also happen to be the seat of consciousness and language—they are the part of our brain that writes the script of our lives and takes all the credit.

Another common intuition is that being rational is a lot of *work*: it requires constant attention and effort. One thing that the traditional metaphors about conscious control get right is the relative strengths involved. Compared to the relatively puny ACC and lateral PFC, the rest of the brain *is* very much like a team of wild horses or a surging river of water. The conscious mind can steer this power in a certain direction, or even completely divert it, but the process takes time and requires a lot of energy. Conscious control is *slow* and *costly*. Experiments in a wide variety of contexts have demonstrated that subjects who are overriding their hot cognition take longer to respond than those who are simply going with their initial reactions. Remember that feeling of *oomph*, and the noticeable pause, that occurred when you were asked to read *LOWER* but say "upper." The times involved are minuscule—we're talking about seconds or fractions of a second—but every little bit counts in the game of survival. An organism that used up precious milliseconds every time it made a decision would soon be left in the dust.

Moreover, work in social psychology has made it clear that cognitive control is a limited resource. When a teacher taps on a dozing student's desk and says, "Pay attention!" it turns out that this is not just a metaphor: attention is costly, and if it is "spent" on one task there is less available to spend on another. This phenomenon is known as "ego depletion." Exerting conscious cognitive control in one domain—say, choosing a healthy radish over chocolate, or suppressing an emotional reaction—makes you subsequently less able to exert it in another domain, like persisting in trying to solve a difficult puzzle. The moral? Effort is effort, mental or physical. Although all brain functions require energy, cognitive control, which requires overriding automatic, deeply rooted actions, is particularly greedy in this regard.

It seems, then, that we have a fundamental problem to solve. Conscious control is crucial for civilized human life. You could never get large numbers of people to live and work together without employing it on a large scale. But this sort of control is physiologically expensive, fundamentally limited in nature, and easily disrupted. Cultural information rides for free in people's memories and in the physical environment around us (marks on paper, tools), but our ability to actually *use* that information is constrained by the choke point of our limited-capacity consciousness.

One of the great strengths of Confucianism—and what differentiates it from recent Western thought—is how it draws upon *wu-wei* to get around this particular problem. As with large wild animals and rivers, the answer lies in *domestication*: channeling the flood waters, or taming the wild animals. Plato's charioteer is always having to pull on the reins because his horses are constantly threatening to run out of control. His job would be much easier if he could get his wild horses to settle down and work together. One of the great facilitators of human civilization is the ability of human beings to consciously shape the behavior of plants, animals, and rivers to better accord with our needs. In the same way, the conscious mind can acquire new, desirable goals and then download them onto the unconscious self, where they can then be turned into habits and implemented without the need for constant monitoring. Effortful, conscious action can be transformed into *wu-wei*.

Consider what happens when we learn a new physical skill, like how to drive. When you're sixteen and out for your first driver's ed class, you really have to focus, taking verbal instructions from the instructor sitting next to you and carefully thinking about how you turn the wheel or press on the pedals. Your prefrontal regions responsible for conscious awareness, especially the lateral PFC, are firing at full capacity, which is why you have to be protected from other distractions, like listening to the radio or talking to the other students in the backseat. (Cell phones were not yet a danger when I learned to drive in the 1980s, struggling to pilot a 1972 Ford LTD the size of a small oil tanker.) At this stage, you're in constant danger of losing control of the car if your attention wavers for even a second.

As you practice, however, this constant, top-down focus is slowly trickling down to other regions of the brain, particularly the *basal ganglia*. The basal ganglia are a set of neuron clusters sitting below the cerebral cortex that seem to be in charge of automatic motor routines—actions that you've learned so well you no longer have to think about them. When we talk about "muscle memory," we should really say, "basal ganglia memory." As we repeat complex motor activities, like driving a car, the basal ganglia and relevant sensory-motor systems are essentially watching and taking notes. Over time, the online control of our actions is shifted to them. In fact, clever experimental manipulations have trained subjects in motor tasks while paralyzing their conscious minds by overloading their working memory with a counting task. Brain imaging shows that, under such conditions, the basal ganglia and relevant motor regions can essentially learn the skill on their own, leaving the subject with no conscious awareness of the new ability. Studies have shown that, as one's skill level increases in a given activity, brain activity gradually drops, and fewer brain regions are involved. As the skill becomes "internalized," the conscious mind can gradually relax and let the basal ganglia and sensorimotor systems take care of things on their own.

TRYING NOT TO TRY: ARTIFICIAL NATURALNESS

Knowing how we in fact master a new skill, like driving a car, gives us an important insight into the *wu-wei* strategy pursued by Confucius and Xunzi. The Confucian Way—the rituals and knowledge passed down from the Zhou dynasty—needs to be fully absorbed by people if they are to live together in a harmonious society. The mind has the power to comprehend the Way, but this is not enough. One needs to get beyond merely understanding the Way and *live* it, in a fully embodied manner. Cold cognition must be made hot. The Confucians tried to accomplish this by intensively training people's embodied minds until consciously learned processes could be performed in an *wu-wei* fashion, like operating a clutch or tying your shoes.

Consider a passage from the *Xunzi* that provides a concise account of how human beings moved from a state of nature to civilized life. The state of nature represents hot cognition gone wild. The results are not pretty, as is evident in any part of the world where institutions have collapsed or states have failed. Luckily, Xunzi explains, the legendary sage kings came riding to the rescue: "The ancient sage kings viewed this chaos with revulsion. Therefore, they instituted rituals and norms of proper behavior, so as to establish social ranks and divide up resources in an orderly fashion. In this way they were able to cultivate and shape people's desires so that they were actually capable of being satisfied. They ensured that desires would be satisfied by the resources at hand and that the resources available in the world would not be completely exhausted by desires." Here we have an interesting sequence of hot-cold-hot cognition. The ancient sage kings were revolted by the chaos of the state of nature. This feeling of disgust is a classic hot reaction. Motivated by this emotion, though, they engaged in some cold cognition. Given the limited resources in the world, and the needs and abilities of various human beings, what would be the best way to distribute them? The solution was the creation of a social hierarchy, maintained by ritual, that grants people access to resources depending upon their perceived worth to society.

Contemporary Western philosophers would now declare Mission Accomplished: we've figured out the problem, we have an answer, all we need to do now is explain it to everyone. Xunzi doesn't stop here, however. For him, the intellectual solution is only the first step. The crucial next stage is to train people to actually alter their *desires* in a way that accords with this solution. Showing a much more sophisticated understanding of how human cognition works than recent Western philosophers, Xunzi argues that acculturation requires a kind of time-delayed cognitive control. The fruits of cold cognition (how to best distribute limited resources) need to be built into hot sensorimotor processes (through ritual and learning) in order to be psychologically effective.

Rituals are essentially behavioral training, encompassing everything from how to conduct important public religious ceremonies to how to dress, enter a room, eat your food, and interact with your parents. Elders and social superiors get the choicest cuts of meat and most comfortable seats, and young people and social inferiors learn to delay their own gratification. Children defer to their parents in all matters, ask about their health a specified number of times each day, and don't travel abroad or make an important decision without formally consulting them. Confucian ritual therefore encompasses both religious ceremony and what we'd tend to think of as good manners. Everyone gets a ritually defined social role or, more properly, a situationally defined *set* of roles: depending on the context, and the person you are interacting with, you may be a public minister, a husband, a father or son, a junior or elder, a teacher or student. Each of these roles comes with a defined repertoire of proper behaviors and a specific set of obligations and privileges. Training in ritual is supplemented by learning, which involves the memorization of classical texts, accompanied by group discussions of their meaning and how they might apply to contemporary life. Learning provides the student with everything from quotations and socially useful turns of phrase to exemplary models of the most important role-specific duties and guidelines for interacting with others.

The sheer scope and detail of early Confucian ritual and learning give it an exotic and archaic feel—it's hard for us to imagine living such a regimented life. Here are a few of the more finicky examples from the *Analects*, describing Confucius's ritually perfect behavior:

In summer, he wore a single layer of linen or hemp, but always put on an outer garment before going out. With a black upper garment he would wear a lambskin robe; with a white upper garment he would wear a fawnskin robe; and with a yellow upper garment he would wear a fox-fur robe. His informal fur robe was long, but the right sleeve was short. He required that his nightgown be knee-length. He wore thick fox and badger furs when at home. Except when he was in mourning, he never went anywhere without having all of his sash ornaments properly displayed.

He would not sit unless his mat was straight.

When mounting his carriage, he would always stand facing it directly while grasping the mounting strap. Once in his carriage, he would not let his gaze wander past the crossbar in front of him or to either side, he would not speak rapidly, nor would he point with his hand.

The idea of regulating one's behavior to this degree seems bizarre and distasteful to many of us. We don't think much about ritual, and tend to think negative things when we do.

Ritual is not, however, as foreign as you might think—it's all around us if we just pay attention. We believe grown children should go their own way but look down upon a child who can't take time out of her busy day to call and wish her mother a happy birthday. We find the elaborateness of ancient dress codes quaint and out-moded but would never show up at an important business meeting in a sweaty jogging outfit. The next time you go out in public, think about how much of what you do and say is scripted: the way you address people on the street and in stores, how you greet a friend versus how you greet a colleague, how you deal with complete strangers as opposed to acquaintances. The social conventions that pervade even our relatively freewheeling modern life might seem insignificant, but try spending a day without them. Go into a store and, instead of making eye contact with the storekeeper and saying hello, just declare what you want to purchase and slap your money down on the counter. You'll make yourself an outcast very quickly and also have an immediate negative impact on the social flow around you. There is considerable evidence that small gestures, tone of voice, and facial expression can change the mood of your social surroundings, with effects that then radiate out in larger circles.

An image that recurs in the writings of Confucius and Xunzi is that of cultural education as a process of carving or reshaping an otherwise poorly constructed self, using ritual and learning as tools. Ritual, for instance, "trims" our inborn emotions, restraining feelings that, by nature, are too extreme. The child's boundless desire to have her needs gratified is transformed into a proper sense of what is reasonable to want and expect. Similarly, emotions that are underdeveloped are enhanced or polished. Our sense of indignation at unfairness is often outweighed by laziness or fear, but the Confucians believed that proper training could strengthen our moral courage. This aspect of Confucian ritual also seems plausible from a contemporary perspective. Work in the science of emotion has demonstrated that social scripts and guided reflection play a crucial role in turning instinctual responses into mature reactions. For instance, babies are born with an innate ability to smile in response to pleasure—a simple reflex reaction. It is not until they are a couple months old that they are able to smile *at* someone, in order to elicit a response or communicate happiness. Later still comes the ability to smile in response to *imagined* happiness (as in a story) or to the happiness of others (empathetic happiness). This process is often encouraged by stories, art, peer modeling, and literature.

The Confucian view of learning also has much to teach us. For instance, we are mostly unaware of the extent to which we rely on stories to help us know how to behave. Of course, those with traditional religious commitments draw on a rich body of paradigmatic tales to reshape their emotions and guide their behavior. For instance, What Would Jesus Do? (WWJD) is an important motto for many in the United States. Even those who are not traditionally religious, however, are constantly drawing upon their own idiosyncratic collections of texts. In trying to teach our daughter how to deal with her emotions and properly interact with others, for instance, my wife and I often turn to Roald Dahl's *Charlie and the Chocolate Factory*. Charlie, a perfectly fleshed-out model of a good-hearted, generous boy, and Veruca Salt, the wonderfully out-of-control, spoiled brat, have proven particularly useful in this regard. The mere word *Veruca!* is usually enough to jar my five-year-old out of obnoxiously self-centered behavior, and we often ask her, when she is making a decision about whether to share or perform another generous act, to reflect on what she thinks Charlie would do. The motto What Would Charlie Do? (WWCD) has yet to attain the popularity of WWJD, but it serves an identical function.

Another way that Confucianism helps to fill in contemporary blind spots has to do with the relationship between learning and thinking. Modern Westerners tend to stress the importance of creativity and thinking for oneself—to borrow from Apple, we all want to *Think Different*. If the Confucians were to adopt a competing marketing motto, it would be *Think Ancient*. They believed that the accumulated wisdom in classics like the *Book of Odes*, an important and sacred collection of poetry and song, should form the very basis of one's mental life. This is why Confucius compares a person who doesn't know the *Odes* to someone "standing with his face to a wall." Moreover, even figuring out the meaning of the *Odes* is a tricky proposition—you can't grasp the meaning of any ancient text by sitting alone in your room. To be properly understood, the Confucian classics need to be studied in a structured social context, under the guidance of a wise instructor and in the company of eager fellow students.

This emphasis on tradition, authority, and collectivism goes very much against Enlightenment thinking. René Descartes, in his *Meditations* (1641), famously declared that you can't accept as reliable knowledge what you've been taught in school. The only way to obtain true knowledge, he argued, is to acquire it yourself, building up logically from first principles. This is a seductive idea, and very deep-seated in our culture, but it's almost certainly wrong. To a degree that qualitatively differentiates us from other animals, we are born incomplete, primed to learn specific types of things from our culture but absolutely dependent upon culture to provide them. We celebrate creativity and novelty but tend not to notice the extent to which any artist or business innovator has been shaped by the ideas and efforts of others, both living and dead. There is a trivial sense in which the technological magic worked by Steve Jobs, for instance, depended on preexisting technologies and ideas. There is a more profound sense in which his creativity could have emerged only from the social and cultural milieu of a particular historical moment in Silicon Valley. In this, as in many other respects, the early Confucian view of learning and thinking is probably much more accurate than the one we've inherited from the European Enlightenment.

One way to look at this privileging of ancient wisdom over individual brainpower is as a response to the dangers of uncontrolled hot cognition and the inadequacies of individual cold cognition. Because we all have limitations, the Confucians put in place a system of environmental buffers to keep us on the straight and narrow Way. We can imagine these as something like the bumpers in bowling-alley gutters at kids' birthday parties: they allow even very young children to experience the satisfaction and excitement of knocking down a few pins and are especially important in the early stages, when kids' physical strength just isn't sufficient to get the ball all the way down the lane without straying into the gutter.

Modern readers of the *Analects*, however, are probably most struck—and most put off—by Confucius's extreme cultural conservatism. As we've seen, he would not sit unless his mat was straight, and he gives some strict advice to a disciple who asks about how to become a gentleman: "Do not look unless it is in accordance with ritual; do not listen unless it is in accordance with ritual; do not speak unless it is in accordance with ritual; do not move unless it is in accordance with ritual." Confucius thought that social harmony depended upon people following specific cultural models in every aspect of life. For instance, he thought the ancient Shao and Wu music represented the best of classical, properly composed music, in contrast to the immoral, seductive popular music of Zheng that was the rage among his contemporaries. The lyrics of the Zheng music—preserved in the *Book of Odes*—are somewhat racy, and although little is known about the exact nature of the music that went along with it, apparently it had a simple but catchy beat, was sung by mixed groups of men and women, and gave rise to sexual improprieties. This should all sound very familiar to concerned parents of any nation or age. The Zheng music was, for Confucius, what Elvis Presley was for my parents' generation, or rap is for our own. Tipper Gore would have slapped a warning sticker on it in a heartbeat.

This cultural conservatism amounts to pervasive and long-term control of all elements of the learner's environment, aimed at guiding his or her behavior and thought into channels approved by the ancients. In our modern world such cultural rigidity, as we might view it, seems antediluvian and foreign. From a contemporary psychological perspective, however, this strategy of cultural immersion looks quite sensible. There is a huge, although now somewhat controversial, literature on the phenomenon of unconscious "priming" effects: altering the behavior of a person by making a word or concept salient in a way that is not even noticed. Subjects who unscrambled a word jumble with words that evoked old people ("Florida," "gray," "wrinkle") walked more slowly when they left the laboratory, and those primed by concepts of politeness waited longer before interrupting. Subjects who unscrambled sentences about helpfulness were more likely to pick up objects dropped by an experimenter. Subjects primed by the social role "professor" performed significantly better on a general knowledge task than nonprimed subjects, while subjects primed with the "soccer hooligan" role performed more poorly. Priming effects extend also to physical actions, which gives us a sense of how and why rituals are so effective. Subjects holding a pen in their teeth in a way that made them simulate a smile rated cartoons as more enjoyable than those who didn't—in other words, faking a smile makes you happier, at least temporarily.

The implications of all of this for Confucian training are obvious. Reading about restrained, elegant exemplars may make you more restrained and elegant. Adopting a humble posture, a respectful demeanor, and a proper sitting position may very well make you more humble, respectful, and proper. The cognitive science of ritual is still in its infancy, but we now have some preliminary experimental evidence that ritual behavior does indeed have an immediate feedback effect on attitudes and emotions and can play an important role in fostering group identity and trust. When intensive, multimedia ritual training—sanctioned by religious authority and practiced for one's entire life—is combined with complete mental immersion in the classics, the result is an incredibly effective program of conceptual and behavioral priming.

So carving and reshaping the self constitute a major part of the Confucian strategy. It might seem strange that people who aspired to effortless ease and unconscious grace thought the way to get it was to be so uptight about etiquette and culture. (In fact, we'll see that the Daoists thought Confucius really went off the rails in this regard.) The Confucian Way is not all about restraint and restriction, however. As any sculptor knows, rough cutting needs to be followed by smoothing and polishing. In terms of Confucian self-cultivation, this finishing work was performed using the tool of communal singing and dancing. Harmonious group movement served as the final step of the carving and reshaping process, producing in the end a perfectly rounded Confucian gentleman.

One early Warring States Confucian text notes that what is special about music is "its ability to enter inside and pluck at the heart-strings." This idea of music as especially capable of "entering" people and transforming their emotions quickly and directly is a theme in many early Chinese texts, as is the idea that the connection goes both ways: listening to someone's music gives you an instant and accurate read of their character. Indeed, the traditional account of why the classic *Book of Odes* contains so many folk songs about mundane topics like farming, weaving, or bucolic love is that the Zhou kings commissioned special ministers to travel around the realm, recording the music and songs being produced by the common people in order to get a sense of their genuine state of mind. Your local governor in province X may tell you that everything is going just great, but if the people working in the fields are singing the blues you might want to look into the situation more carefully. (The modern equivalent of this is the daily summary of Chinese microblogging sites provided to the ruling elite.)

The idea of diagnosing popular sentiment by listening to popular music is not that crazy. Experimental work suggests that people are quite accurate at identifying certain emotions—like anger, sadness, and happiness—motivating a piece of music, even when the music comes from a completely unfamiliar culture. The effect also seems to flow in the other direction, something we are all familiar with intuitively: listening to angry music makes you angry, and sad music makes you want to curl up with a bottle of whisky and ruminate on your life's failures. This is why huge, aggressive marching bands pump up football players headed onto the field, and why happy dance music is people's first choice for celebrating marriages or other auspicious events.

Another important thing to note about the social use of music is that it is almost always *communal* and accompanied by dance or synchronized movement. This shared, physical activity seems effective not only in conveying specific emotions but also in bringing people together into bonded communities. Western scholars of religion such as Émile Durkheim (1858–1917) have long theorized that the social cohesion created by music and dance is the very reason they are so common in religious traditions around the world. Ritualistic drills and chanting have been used throughout history to both create and express solidarity. My personal favorite example is the haka, a traditional Maori war dance used by New Zealand's national rugby team, the All Blacks, before matches. Check it out, if you dare (type "New Zealand All Blacks haka" into YouTube). This beautiful distillation of righteous anger and coiled violence not only brings the All Blacks together as a team and infuses them with energy and motivation but also scares the hell out of the opposing side.

Music goes both ways, both creating new emotions in listeners and serving as evidence that one is feeling the way one should. Music and dance appear to bypass the cold, and often sneaky, conscious mind in order to get right at our embodied hot cognition. The science is still in its infancy, but some recent experimental work supports the theories of both the ancient Confucians and more contemporary scholars of religion: synchronized group activities like music, dance, and marching help communicate emotions, enhance group identity, and increase interpersonal cooperation, even when the subjects are randomly selected university undergrads, complete strangers thrown together for a single experiment in the artificial and stimulus-poor environment of the lab. It is this sort of ecstasy—literally, *ec-stasis*, or being outside oneself—that the Confucians hoped would allow aspiring gentlemen to make the transition from disciplined self-control to spontaneous, but properly ordered, joy in the Way. The rapture inspired by sublimely beautiful music represents, in the Confucian view, the end of a long journey of self-cultivation, where one arrives at an entirely new place that nonetheless feels completely familiar.

CONFUCIAN *WU-WEI*: AT HOME IN CIVILIZATION

Despite all the cultural conservatism, then, the Confucian view of civilized life is, in the end, optimistic. Sigmund Freud, to take a prominent Western counterexample, saw the tension between hot and cold cognition as the ineradicable tragedy at the heart of modern life. We were miserable in the state of nature because a world in which everyone allowed their id—their hot cognition—to run wild would be chaotic, capricious, and brutish, even for the rare individuals lucky enough to be at the top of the social pyramid. A civilized life is better for everyone overall, but it also exacts a cost: everyone is required to either repress or sublimate a large portion of their instinctual drives and to live under the iron rule of cold cognition. The result is a state of what Freud calls *Unbehagen*, which is usually translated as "discontent" or "dissatisfaction" but also includes a sense of physical unease. If you go into a weird, creepy old house and feel the hair stand up on the back of your neck at the thought of ghosts, that's also *Unbehagen*. For Freud, to live as a civilized human being is to live in a creepy old house.

The Confucian ideal of the perfectly civilized person is thus the opposite of the *Unbehagen*-ridden Freud. The Confucian strategy does try to boost self-control, but mainly it tries to make cold-cognition–approved behavior more reliable by making it "natural" and spontaneous. The Confucian gentleman is perfectly at home in civilization, instead of always uneasy about it. An evocative passage from the *Xunzi* describes how Confucian learning can enter the aspiring Confucian and take control of his body, changing his very physiological makeup. "The learning of the gentleman enters his ear, becomes firm in his heart-mind, spreads out through his four limbs, and manifests itself in both activity and repose. In his merest word, in his slightest movement, the gentleman can always be taken as an example and a model." At this point there is no more need for cognitive control. Every action of the gentleman is free and easy, yet perfectly correct. "The sage gives free rein to his desires and fulfills all of his emotions, but having been regulated they accord with civilized norms," Xunzi explains. "What need has he, then, for strength of will, endurance, or cautiousness?" The idea is that, for someone like Confucius at age seventy, all that uptight ancient cultural stuff has become like the water in which a fish swims, completely unnoticed and yet perfectly comfortable.

This is Confucian *wu-wei*. The idea of replacing problematic, innate dispositions with new, socially desirable ones seems like the perfect solution to the problem of how to get people cooperating in large societies, and in the case of Confucius himself it seems to have worked like a charm. At age seventy, he serves as a model of the ideally socialized person: elegant, perfect in etiquette, yet completely at ease and sincere. Lurking in the background of early Confucian writings, however, we find hints that self-cultivation doesn't always work out so well. Indeed, we are at times given the impression that, for every effortlessly graceful and sincere Confucius, the hard grind of self-cultivation produces a dozen wannabes, counterfeits of virtue who talk the talk, and even walk the walk, but lack the inner commitment of the true gentleman.

BEWARE THE VILLAGE POSEUR

Why is spontaneity something we have to cultivate? In other words, why do we have to try not to try? Confucius and Xunzi have a clear answer. We're born with some deep flaws, so just going with our instincts will lead to nothing but disaster. And they are no doubt right about this. For reasons we will explore in more detail, it is likely that the set of emotions and desires with which we're born—our innate values, if you will—are ill-suited to the complicated social world that we all inhabit. There is a reason that children need to be educated. They need to learn technical skills, of course, like math, handwriting, geography. It's more important, though, that they learn to *behave*, and this involves valuing things—other people's needs, playing by the rules—that children tend not to value spontaneously. This means that any *wu-wei* worth having is going to be the product of an unconscious mind that has been guided and shaped by conscious design and instruction.

Unlike the honing of physical skills, however, the inculcation of Confucian character virtues presents a conundrum. The early Confucian texts repeatedly emphasize, for instance, that you can't become a real Confucian unless you, at some level, already *love* the

Confucian Way. As Confucius warns his students: "I will not open the door for a mind that is not already striving to understand, nor will I provide words for a tongue that is not already struggling to speak. If I hold up one corner of a problem, and the student cannot come back to me with the other three, I will not attempt to instruct him again." The point is that the Master cannot teach someone who is not driven by a need to learn, and he cannot impart the Way to someone who does not want it. Education is effective only when there is active, sincere, and appreciative engagement on the part of the learner.

For Confucius, the problem with his contemporaries is that they *don't* have this kind of appreciation. They value all the wrong things—fame, money, power, sex, food—when what they really should value and love is the Confucian Way. As he remarks rather sharply at one point, "I have yet to meet a man who loves *de* as much as the pleasures of the flesh." In another passage, Confucius complains, "Is goodness really so far away? If I merely desire goodness, I will find that it is already here." A healthy bit of frustration shows through in these passages. We all have the ability to be good if we would simply desire it as we should, but how can one instill this desire in someone who does not already have it (or who desires the wrong things)? We see this problem at work when a disappointing disciple explains to Confucius that he does, really, love the Way but feels that he just can't find the strength to pursue it. Confucius slaps him down: "Those for whom it is genuinely a problem of insufficient strength end up collapsing somewhere along the Way. As for you, you deliberately draw the line." This is the heart of a paradox that Confucius faced, a manifestation of the "paradox of *wu-wei*" described earlier.

How can you consciously develop a new, but completely sincere, desire or emotion that you don't already have? What happens if you study the Way of the Zhou without the right motivation? That is, what if you practice all of the rituals that Confucius says you should practice, you learn the *Odes*, you sing the right songs and dance the right dance, but you don't have that special inner love? The answer is that you become a *counterfeit* of a good person: someone who looks like a good person on the outside but has no real virtue on the inside. At one point Confucius talks about the "village poseur," whom he characterizes as a "thief of *de*." One of the best explanations of what he meant by that is provided by his later follower Mencius:

> Those who try to censure the village poseur can find no basis; those who try to criticize him can find no faults. He follows along with all the popular trends and harmonizes with the sordid age. Dwelling in this way he seems dutiful and trustworthy; acting in this way, he seems honest and pure. The multitude are all pleased with him—he is pleased with himself as well!—and yet you cannot really discuss with him the Way of the ancient sage kings. This is why he is called the "thief of *de*." Confucius said, "I despise that which seems to be but in fact is not. I despise weeds, for fear they will be mistaken for domesticated sprouts. I despise glibness, for fear it will be mistaken for rightness. I despise cleverness of speech, for fear it will be mistaken for trustworthiness. I despise the tunes of Zheng, for fear they will be mistaken for true music. . . . I despise the village poseur, for fear that he will be mistaken for one who truly possesses *de*."

By serving as a counterfeit model of virtue for the common people, the village poseur is in effect a false prophet, not only blocking the development of true virtue in himself but also leading others astray. This is why Confucius hates him.

It's clear that Confucius feels that most of his contemporaries who claim to be scholars and gentlemen are really just such poseurs, mechanically acting out rituals and memorizing texts in a way that is ultimately hollow and meaningless. At one point a disciple asks about filial piety, or proper behavior toward one's parents. The Master replies, "Nowadays 'filial' means simply being able to provide one's parents with nourishment. But even dogs and horses are provided with nourishment. If you are not respectful, wherein lies the difference?" To be a filial child, it is certainly the case that you have to perform certain physical actions. You need to visit your parents regularly and make sure they are comfortable, properly fed, and in good health. This is, however, not enough—as Confucius says, we do as much for our domestic animals. True filiality involves doing all of these actions with the right *attitude*, one of love and respect, and it's clear that just going through the motions for a long time is not sufficient to instill this attitude.

So where does that leave us? In the end, Confucius's strategy seems to be an injunction to just keep plugging away: try to pay attention and see the point, and eventually the love will come. Xunzi makes this even more explicit—he doesn't even require the initial enthusiasm. Just grind away, he says, and love for learning will be born, trust me. Really. The danger is that the grinding-away strategy is just going to result in a world full of hypocritical poseurs, pretending to be committed to public values but really just after their own private gain. A contemporary equivalent would be a dating scene where everyone is pretending to be someone they are not, and to value things that they do not actually value, where the only sort of connection is ultimately a robotic charade: hollow people with faked interests and emotions playing out a script written by someone else, the only genuine motivation being a desire for self-satisfaction. Depressing, no?

Now you have a sense of the motivation behind those who, presented with the Confucian vision, recoiled in horror and ran for the woods. The Confucian strategy for attaining *wu-wei* needed to be laid out in some detail, not only because it was the dominant one in early China but also because it's the strategy that every other thinker had to grapple with. One could argue that it's the basic strategy that *anyone* concerned with naturalness must begin with, since—despite the rhetorical stance taken by some of the Daoists—we are neither fish nor birds. Human beings are above all cultural animals, and that means that any type of *wu-wei* worth wanting is going to require education and effort. Having conceded these points, one must also note that there are serious tensions inherent to the Confucian approach, and no one was as eager to point them out as an early Chinese counterculture figure known as Laozi.

4

Stop Trying

EMBRACING THE UNCARVED BLOCK

THE *ANALECTS OF CONFUCIUS* WAS PUT TOGETHER BY HIS DISCIples after he died and consists of multiple chronological layers. The later sections, probably composed considerably after Confucius's death, contain a small number of odd stories where Confucius is shown confronting people who seem less than enamored with his project of arduous cultural training as a means of achieving *wu-wei*. Confucius, of course, triumphs over them and proves the superiority of the Confucian Way, but we are getting here our first glimpse of a countermovement that would grow into an important strand of Chinese religion.

In one of these stories, Confucius and his entourage are on the road and find their way blocked by a river. Unfamiliar with the region, Confucius spots two men in a nearby field, yoked together and pulling a plow. This is strange. In Confucius's time plows were typically drawn by oxen, so it's surprising to see people yoked to a plow themselves. This is the early Chinese equivalent of, say, someone in suburban New Jersey using a hand-pushed mower instead of a power mower to cut his front lawn. These guys are deliberately using a primitive and outmoded technology in order to make some kind of point. Things quickly get weirder.

Confucius, mounted on a chariot, sends one of his disciples, Zilu, to ask these two plow-pulling guys where the best place to ford the river might be. When Zilu approaches and addresses them, we learn that their names are as bizarre as their behavior: Standing Tall in the Marsh and Prominent in the Mud (I'm not making this up). We can assume that these are not the names given to them by their parents, so we have here yet another puzzling element: people who have rejected their proper given names—shocking in a traditional Confucian context—and have adopted metaphorical and vaguely rebellious monikers instead.

Their response is, if anything, even odder. Having been formally approached by Zilu (always a stickler for protocol) and politely asked about the location of the ford, Mr. Standing Tall in the Marsh ignores the question and instead rudely demands, "That fellow holding the reins there—who is he?" The disciple, taken aback, meekly informs him that it is Confucius. "Do you mean Confucius of Lu?" Yes. "Then *he* should know where the ford is." The two men proceed to ignore Zilu and start pulling their plow again. Their jab probably refers to the fact that Confucius spent his entire adult life traveling around China trying to get some ruler to listen to him and put the Confucian Way into practice. Someone who spends that much time on the road should know his way around. Behind the shocking rudeness, then, we can perceive some knowledge of who Confucius is and what he is up to. These are no ordinary plowmen.

In addition to being extremely polite, Zilu is also rather pigheaded, so he gives it another try. This time he addresses Mr. Prominent in the Mud, who instead of answering responds with another brusque question, "Who are you?"

"I am Zilu."

"The disciple of Confucius of Lu?"

"Yes."

"The whole world is as if engulfed in a great flood, and who can change it? Given this, instead of following a scholar who merely avoids the bad people of this age, wouldn't it be better for you to follow thinkers like us, who avoid the age itself?"

The two then turn their backs on Zilu and go back to work. Who are these guys? They describe themselves as "scholars," like Confucius, but they live in the countryside, working the fields with their own hands, using deliberately primitive technology. They have rejected not only their proper given names but the very age into which they've been born.

This encounter with the odd plowmen recalls another, similarly strange, passage in the *Analects*. There we find Confucius in the state of Wei, where he is living temporarily while he tries to win over the local ruler. It isn't going well. The passage opens with Confucius sitting in his doorway, playing a slow, mournful tune on the stone chimes, the early Chinese equivalent of strumming an acoustic guitar and singing the blues. A man with a wicker basket strapped to his back—typical peasant garb—is passing by and pauses to listen. After a few minutes, he remarks, "Those chimes are being played by someone with a lot on his mind!" After listening for a bit longer, he adds, "What a stubborn, petty attitude! If no one understands you, just tend to yourself!" Mysterious basket man then quotes a stanza from the *Book of Odes* ("If the river ford is deep, use the stepping-stones / If it is shallow, simply raise your hem") and heads off on his way, presumably to join his friends Standing Tall in the Marsh and Prominent in the Mud.

Though dressed like a peasant, this guy has an ear for music and can quote from the *Odes*. Like the rude plowmen, he is clearly no ordinary commoner but rather a classically trained scholar, a member of the elite, who has gone into voluntary reclusion. The plowmen describe themselves as "fleeing the age," and the music critic's apropos citation from the *Odes* seems to provide the rationale: you need to go with the flow. If the world has gone to hell, trying to change it will only make it worse. The tune that Confucius was playing must have been intended by him to convey a feeling of quiet commitment and frustrated purpose. To his critic's ear, however, it feels more like ignorant recalcitrance. If Confucius is having no success, he should not force it. He needs to stop trying to shove the Confucian Way down everyone's throats and instead tend to his own garden. Reject technology, quit your job, ditch your fancy clothes, and head off to the countryside.

If this is beginning to sound a lot like the 1960s, it shouldn't be surprising. Swap out Standing Tall in the Marsh for Moonchild, a human-drawn plow for a homemade composting toilet, and a facility with the *Odes* for an ability to quote Emerson and Thoreau, and these recluses from the later strata of the *Analects* would find themselves right at home in 1968 Haight-Ashbury. They are, as far as we know, the *original* hippies, dropping out, turning on, and stickin' it to the Man more than two thousand years before the invention of tie-dye and the Grateful Dead.

These hippies don't fare all that well in the *Analects* itself—as one might expect, in a book compiled by Confucius's followers, Confucius always gets the last word. All of the "primitivist" encounters in the *Analects* end with Confucius justifying his refusal to take the easy way out, and with the recluses being made to look like lazy dropouts. After the encounter with the plowmen, for instance, Confucius sniffs, "A person cannot flock together with the birds and the beasts" and characterizes the lifestyle of these recluses as an irresponsible rejection of a gentleman's responsibility toward himself and his society. This should also sound familiar, much like the Eisenhower generation's reaction to their children dropping out of school and giving up bathing. In any case, the hippies soon get their *own* book, where they are able to convey their message directly, free of Confucian spin-doctoring.

This book has come down to us with the title *Daodejing*—literally, *The Classic of the Way and Virtue [de]*—and is probably one of our oldest transmitted Warring States texts after the *Analects*.

It has traditionally been attributed to a mysterious sage known as Laozi (Lao-Tzu), or "The Old Master," and is therefore often just called the *Laozi*. "The Old Master" is precisely the sort of made-up name one would concoct to give additional gravitas to one's writings, and scholars have long assumed that the text of the *Laozi* is the product of many hands, probably just one of many compilations of primitivist writings to survive. I'll nonetheless follow convention by using "Laozi" as a useful shorthand for whoever compiled the standard received version of the text.

Like the recluses we glimpse in the later sections of the *Analects*, Laozi saw himself as living in a profoundly corrupt age, characterized by glaring social inequities, economic chaos, and superficial consumerism. In a passage that sounds as if it could have been lifted from a contemporary "Occupy Wall Street" pamphlet, he complains:

> The court is corrupt,
> The fields are overgrown,
> The granaries are exhausted.
> And yet some wear clothes with fancy designs and colors,
> Hang sharp swords from their belts,
> Stuff their bellies with fine food and drink,
> And possess more wealth than they need.
> This is what is called "being proud of being a robber."
> Far is this from the Way!

Like Confucius, Laozi saw the corruption around him as a sign that the world was not in accord with the Way of Heaven, and his goal was to bring the two together again into *wu-wei* harmony. His view of the Way, however, bears no resemblance to that of Confucius and in fact was formulated in direct opposition to it.

DOWN WITH THE MAN
(AND MADISON AVENUE): SOCIAL KNOWLEDGE
AND THE HEDONIC TREADMILL

To begin with, unlike Confucius, Laozi thought that *less*, rather than more, culture was the answer. Confucius saw humans as fundamentally crude by nature and in need of cultivation. Laozi believed that we're fine as we are—or at least as we *were*, before society messed us up. For Laozi, human nature is fundamentally good, and our innate dispositions are the ones we need to follow. Education and training are therefore entirely counterproductive, leading us away from our essential goodness. Confucius saw human taste as something to be slowly refined over time, much as a connoisseur acquires an ever more subtle appreciation of increasingly complex wines. Laozi saw acquired tastes and cultural innovations as a source of disorder:

> The five colors blind our eyes.
> The five notes deafen our ears.
> The five flavors deaden our palates.
> The chase and the hunt madden our hearts.
> The pursuit of goods hard to come by impedes our way forward.
> This is why the sage is for the belly and not for the eye;
> He casts off the one and takes up the other.

This contrast between "the belly" and "the eye" makes for a wonderful metaphor. The belly is the seat of our basic desires, which in Laozi's view are quite modest: some plain food to eat, water to drink, a roof over the head, maybe a little low-key fun in the sack, and *basta*. The eye, on the other hand, is constantly getting us into trouble, because with our eyes we can see things that are far away, things that we don't have but want as soon as we spot them. In the distance we can see other foods that look more delicious than ours, shiny objects that draw us closer, women and men who are younger and more attractive than our current partners.

Moreover, the very act of socially labeling things as "beautiful" or "good," in Laozi's view, already distorts our natural, spontaneous judgment. One passage tells us that "when everyone in the world knows that the beautiful is 'beautiful,' it is then ugly." What this cryptic claim might mean is spelled out by the rest of the passage, which argues that singling out and verbally labeling a value sets up a dichotomy that then acts as a mental trap. Laozi famously declared, "He who speaks does not know," a motto that underlines the tendency of verbal labels to cloud our judgment and our ability to see what is right in front of us. The philosophical target, of course, is the Confucians, with their clear standards concerning the right kind of music to listen to, the right kind of clothes to wear, the right way to enter a room, and—perhaps most damaging—the precise way to be "good." Laozi's argument is that calling some behavior "good" ensures that it will *not* be good, because conscious labeling and explicit effort poison our experience.

The idea that attaching verbal labels to things can alienate us from our own experience finds some confirmation in contemporary psychological research. A substantial literature on "verbal overshadowing," for instance, suggests that consciously reflecting on our perceptions or evaluations of taste, and then being forced to put them into words, actually impairs our judgment. In one classic experiment, Tim Wilson and Jonathan Schooler had subjects taste five different types of jam, and asked some of them, as they tasted, to explicitly note the qualities that they liked and disliked and to reflect carefully on their reasons for preferring one jam over another. In the control condition, the subjects just sampled the jams and rated them. Wilson and Schooler found that the process of analysis not only altered subjects' judgments considerably but altered them in a negative direction: the choices of the analyzers were worse than those of the nonanalyzers, as measured against expert judgments of jam quality. A similar study involved the evaluation of dorm-room posters, which the subjects were allowed to take home with them. The researchers found that reflective analyzers made poor choices, even measured against their own long-term evaluations, because several weeks later they reported less satisfaction with their choices than nonanalyzers. They grew to hate that Monet poster they'd so eloquently praised during the experiment. This work is part of what is now a huge literature on the often harmful effects of rumination and explicit analysis on people's ability to experience and identify pleasure.

In Laozi's view, two of the main forces leading us away from *wu-wei* are the negative effect that thinking and verbalizing have on our ability to simply experience life, combined with the tendency of our desires to grow incessantly, becoming temporarily sated but then aroused again by some more desirable mirage in the distance. As he argues in one passage,

> There is no crime greater than indulging your desires;
> There is no disaster greater than not knowing contentment;
> There is no calamity more serious than wanting to get ahead.
> If you can know the contentment of contentment, you will be
>   forever content.

"Knowing the contentment of contentment" requires resisting the siren call of consumer culture and instead holding fast to primitive and simple pleasures. Again, from a contemporary perspective, there is a lot to be said for this insight. The desires of the eye form, of course, the entire basis of the modern advertising industry, which has turned the continuous ramping up of our desire for "goods hard to come by" into a refined science. The minute the latest iPhone is released, our current iPhone suddenly seems less attractive; the advent of a new fashion trend in Paris suddenly renders shipping containers full of otherwise perfectly good clothes almost worthless. As Henry David Thoreau once complained, "The head monkey at Paris puts on a traveller's cap, and all the monkeys in America do the same"—just one of the many features of civilized life that drove him to the shores of Walden Pond. The darker side of this distorting effect of artificial social norms can be seen in the unhealthy body images that dominate modern advertising, where seventeen-year-old anorexics are held up as the model of female beauty. Men are trained that this is what they should find attractive, and too many women harm themselves trying to conform to the standard. Laozi would see this as a perfect example of perverted cultural norms destroying our own natural tastes.

Laozi is right about much of this, but we don't want to jump too quickly to blame Confucianism or Madison Avenue for our restlessness and dissatisfaction. Constant turnover of preferences and feverish striving are not just products of evil marketing men or unrestrained capitalism; rather, they reflect a fundamental aspect of human psychology. Because of the way that we are built, perfect happiness or pleasure seems structurally impossible to attain, at least when pursued in the normal ways. Classic work in the 1970s established the idea of a "hedonic treadmill," according to which positive or negative events result in only temporary increases in happiness or unhappiness. Famously, these researchers claimed to have found that people who had won huge lotteries or (at the other extreme) had became paraplegic as the result of accidents initially experienced intense joy or sorrow but soon reverted to a baseline level of happiness. Follow-up research confirmed that, across a variety of dimensions and life situations, the effect on happiness of events that one would intuitively think would be dramatic and permanent— suffering spinal cord injuries, getting married, suffering the death of a spouse—are in fact often surprisingly transitory.

The basic mechanism seems to be *adaptation*, a phenomenon well-known to perception researchers: after perceiving something for a certain period of time, your sensory system "adapts" to it, causing it to recede into the background. Sensory adaptation is crucial for perception to work, ensuring that our senses aren't completely swamped by myriad stimuli that constantly assault us. This allows novel stimuli to be perceived against a constant background and alerts us to changes in the environment. The hedonic treadmill studies suggest that, in the same way that we grow used to the sounds of a busy city, positive or negative changes to our life situation make an initial mark but then fade into the background.

Another source of dissatisfaction is our incessant need to measure our achievements against those of our peers. Psychologists used to be puzzled about the fact that rising GDP in the West was not resulting in fundamental changes in people's reported levels of happiness. What they found was that, once a certain minimum threshold of material well-being is reached, our objective level of wealth seems much less important than our relative wealth—that is, how we stack up against our neighbors or colleagues. Once you have enough money to buy the basics and indulge in some pleasures, like eating out or buying new clothes, ranked *status* comes to matter much more than wealth per se. Status, in turn, is inherently unstable because it's by its very nature relative—the benchmark is always moving as others around us rise or fall. Moreover, we seem designed to focus more on what we *don't* have than what we do: we are much more irked about those two people ahead of us than pleased about the twenty behind.

There is a very good evolutionary reason for all of this. We and our closest primate relatives are intensely social animals embedded in clear hierarchies, and precisely where we fall in these hierarchies can determine whether we spread lots of our genes or none at all. Individuals who stewed about being number 5, and who stayed up nights scheming about how to surpass or knock off numbers 1 through 4, generally passed on more genes than those who simply kicked back and enjoyed what life had given them. The kind of general social affluence and security that allows the Dude of *The Big Lebowski* (1998) fame to enjoy his long baths, White Russians, and bowling league in peace is a very recent development—and even the Dude can't, in the end, avoid having to confront more harsh realities. The driving force of constantly ramped-up desire and social status seeking is one of the many ways in which psychological adaptations designed to maximize reproductive success come into direct conflict with individual happiness.

So Laozi appears to be wrong that our spiritual ills can be blamed entirely on the Confucians or advertisers. It is reasonable, however, to follow Laozi in seeing our endless peer comparison as the fault of "the eye," rather than the belly, and it extends to many things besides income. Our belly may be perfectly comfortable in our current car, but our eye can see the nicer, newer car in the driveway next door (or in the magazine ad or billboard), and this perception immediately decreases our satisfaction with what we're currently driving. The car itself has not changed the slightest bit, but our benchmarking mind has demoted it anyway. The psychological literature on these topics reads like a footnote to the *Laozi*, a scientific documentation of a phenomenon that he perceived with incredible accuracy. In fact, he even coined a technical term for it, *fan* 反, usually translated as "reversion" or "turning back," but which could just as well be rendered "treadmill." Anything taken to its extreme turns into its opposite, an idea captured at one point in the text by the image of an ancient and famous "tilting vessel" said to have been designed so that it stood upright when empty but tipped and spilled out its contents when filled:

> Grasping the vessel and filling it to the rim
> Is not as good as stopping in time.
> Sharpen the blade, and the edge will soon be dulled;
> When gold and jade fill the room, they will soon be stolen;
> When wealth and honor lead to arrogance, disaster naturally
> follows.

Another image that captures the idea of the treadmill even more succinctly is the famous Yin-Yang symbol:

In this image, which we have to picture spinning slowly counterclockwise, the white *yang* (the strong, bright, male principle) is waxing in power, gradually reaching its apex at twelve o'clock. At this point, the seed of the black *yin* (the weak, dark, female principle) is born and takes over, similarly growing in strength until it reaches full power at six o'clock, when the seed of *yang* now reasserts itself. This *yin-yang* circle is looked upon nowadays as a positive image of mystical wisdom, happily slapped on surfboards and tattooed onto twenty-year-old butts. In fact, it actually symbolizes a dark and pessimistic vision, akin to the Buddhist teaching of *dukkha* or impermanence: all striving leads to disappointment in the end because there is no permanence in the world. The cycle of *yin-yang* is not to be celebrated but escaped.

How do we escape it? By doing nothing. Of all the thinkers in early China, Laozi uses the term *wu-wei* in a sense closest to its literal meaning of "no doing." In his mind, the energy driving the pernicious cycle of reversion is desire—another parallel to Buddhism. Desire, in turn, is created and strengthened by the two activities so fundamental to the Confucian pursuit of *wu-wei*: cultural knowledge and active striving. Cultured tastes and artificial needs, he argues, distort our natural enjoyment of the world, and we can counter this by keeping our food simple and our physical desires modest. The more insidious danger, though, is the Confucian goal of training people to be virtuous. For Laozi, *trying* not to try is not only self-defeating but the source of all human suffering.

GRASP IT AND YOU WILL LOSE IT

The clearest account of how, in Laozi's view, conscious effort and striving are responsible for the world's ills is a famous passage that appears halfway through our received version of the text:

The highest Virtue [de] does not try to be virtuous, and so really
possesses Virtue.
The worst kind of Virtue never stops striving for Virtue, and so
never achieves Virtue.
The person of highest Virtue does not act [wu-wei] and does not
reflect upon what he is doing.
The person of highest benevolence acts, but does not reflect;
The person of highest righteousness acts, and is full of self-
consciousness.
The person of highest ritual propriety acts and, when people do
not respond, rolls up his sleeves and forces them to respond.

Hence when the Way was lost there arose Virtue;
When Virtue was lost there arose benevolence;
When benevolence was lost there arose righteousness;
When righteousness was lost there arose the Confucian rituals.
Ritual is the wearing thin of loyalty and trust,
And the beginning of disorder.

This is the Laozian version of the fall from grace, a description of
a continual decline from an original state of unselfconscious, *wu-
wei* perfection. Once upon a time, people were genuinely virtuous—
they had real Virtue with a capital "V" (*de*), precisely because they
didn't try. Things then got progressively worse. First, we had the
"benevolent" person, who was relatively sincere but still felt the need
to act in the world, to "do good." That's where the trouble started.
Then we had the dreaded "righteous" person, a sanctimonious
poseur who couldn't stop dwelling on how wonderful he was and
felt the need to stick his nose into everyone's business. Even worse
were those who knew nothing but rigid adherence to ritual guidelines, driven by an insatiable urge to see their sense of what was right
imposed on everyone around them. This, in turn, forced people to
become hypocritical—encouraging them to substitute empty forms

of respect for genuine reverence and flowery protestations of love for true affection. For Laozi, this triumph of image over substance is like the rosy glow of a tuberculosis patient, a misleading outward symptom of a deeply entrenched sickness:

> It is only when the great Way falls into disuse that people start talking about "benevolence" and "righteousness";
> Once "knowledge" and "wisdom" emerge, the great hypocrisy begins;
> It is only when family members do not get along that we have talk of "filial piety" and "parental affection";
> It is only when the state is in darkness and chaos that "upright officials" appear.

To be fair to Confucius, this is not what he wanted. He, too, worried about the "village poseur" who goes through all the motions of being good but is in the end a hollow counterfeit of virtue. Laozi's point, though, is that the Confucian scheme—with its emphasis on striving and learning and endless self-reflection—is incapable of producing anything *other* than village poseurs. The very act of trying to be good fatally contaminates the goal. As he says, "The worst kind of Virtue never stops striving for Virtue, and so never achieves Virtue." Conscious, goal-oriented attempts to be virtuous will never lead to anything other than hypocrisy.

The psychologist Daniel Wegner has spent a good deal of his career exploring what he terms the "ironic effects" of conscious, intentional effort. In words that sound like they could have come straight from the *Laozi* he observes, "Many of our favorite goals, when pursued consciously, can be undermined by distractions and stressors to yield not just failure of goal achievement but the ironic opposite of that attainment. We achieve exactly what we most desired not to do." He and his colleagues have amassed a large body of evidence suggesting that we get depressed when we're consciously trying to be happy, anxious when we are trying to relax, and dis-

tracted when we're trying to concentrate. When we try to consciously forget something, we remember it more clearly; when we
try to make ourselves sleep, it makes our insomnia worse. Trying to
stop thinking of sex is the best way to think of sex. If you want to
make someone overshoot when they are attempting a putt in golf,
tell them to try as hard as they can to *not* overshoot. In a wonderful study entitled "The Putt and the Pendulum: Ironic Effects of the
Mental Control of Action," Wegner and colleagues not only documented the pernicious effect of golfing instructions but also showed
that asking someone *not* to move a pendulum suspended from their
hand causes them to move it even more, and specifically in the direction they were told to avoid. The effect was magnified when they
were put under cognitive load (asked to count backwards from one
thousand in intervals of three), precisely the sort of situation your
average Confucian disciple would find himself in, with a whole pile
of classics to memorize and an enormous set of rituals to remember.

These paradoxical effects of conscious effort were drawn upon
by the Austrian psychiatrist Viktor Frankl, who pioneered the use
of "paradoxical intention" therapy. Focusing on the specific problems of insomniacs, for instance, Frankl noted that "sleep [is like] a
dove which has landed near one's hand and stays there as long as one
does not pay any attention to it; if one attempts to grab it, it quickly
flies away." His recommended strategy, therefore, was to advise insomniacs to try to *stay awake* as long as possible; the result is that
they actually fall asleep faster. The famous—or notorious—tennis
player John McEnroe, though not to my knowledge formally trained
as a Franklian psychoanalyst, intuitively perceived the power of this
technique. When faced with an opponent whose forehand, for instance, was working smoothly and perfectly, McEnroe would supposedly compliment him on it as they changed sides: "Wow, your
forehand is really great today." His opponent would then, of course,
suddenly start botching easy shots in the next set.

Ironic effects are not confined to physical movements, emotions,
or physiological processes like trying to fall asleep—they extend to

the moral realm as well. Subjects who are explicitly instructed to be fair and unprejudiced actually become *more* prejudiced. Similarly, the phenomenon of "moral licensing" refers to the *negative* effect that positive moral self-evaluations have on a person's subsequent behavior. Subjects allowed to establish that they are not prejudiced are more likely to exhibit prejudice in follow-up exercises. People asked to imagine that they had just donated time to community service showed a marked uptick in self-evaluation (they were more likely to assent to the statement, "I am a compassionate person"), and were subsequently more likely to splurge on a frivolous luxury item. In a nice study targeted specifically at the pernicious effect of *labeling* morally positive behavior, and thereby puffing up people's conceptions of themselves, subjects in a variety of conditions were asked if they would like to donate up to $10 to a charity of their choice. Those in the neutral condition gave an average of $2.71, while those who were primed with words like *caring, generous,* or *kind* offered a measly $1.07.

Thinking that you are good can make you bad. Talking about positive behavior can encourage negative behavior. Laozi is clearly on to something when he warns us that consciously trying to be righteous will, in fact, turn us into insufferable hypocrites and that anyone striving to attain virtue is destined to fail. What are we then to do? According to Laozi, we just need to go home.

RETURN HOME, EMBRACE THE UNCARVED BLOCK

One of the most striking features of the *Laozi* is the manner in which the metaphors it employs seem directly targeted at the metaphors that dominate the *Analects*. Confucius urges us to adorn ourselves with culture (*wen*, 文), the original meaning of which is "pattern" or "decoration"; Laozi tells us to exhibit the "unadorned" and "simple." Confucius advocates carving and polishing ourselves the way we would a fine piece of jade; Laozi tells us to embrace the "uncarved block." Confucius compares the process of self-cultivation to a grueling, lifelong journey toward sagehood; Laozi tell us to stop, turn back, and go home to our primordial roots, to "being an infant," or to "honoring the Mother." Acutely aware of the potentially ironic outcomes of striving, as well as the pernicious effects of excessive self-regard, Laozi thinks that the only way to attain *wu-wei* is to try to mitigate the damage inflicted by the hyperactive do-gooders and sanctimonious moralists produced by Confucian teachings:

> Get rid of wisdom, abandon distinctions
> and the people will benefit a hundred-fold
> Get rid of cleverness, abandon profit,
> and the people will return to being filial and kind.
> Get rid of artifice, abandon reflection,
> and robbers and thieves will disappear . . .
> Exhibit the unadorned and embrace the uncarved block,
> Reduce selfishness and make few your desires,
> Get rid of learning and you will have no worries.

To attain Laozian *wu-wei*, you need to *undo* rather than *do*, gradually unwinding your mind and body, shedding book learning and artificial desires. The goal is to relax into a state of perfect nondoing (*wu-wei*) and unselfconsciousness, like settling into a nice warm bath:

> One who engages in learning adds to himself day by day;
> One who has heard the Way takes away from himself day by day.
> He takes away and then takes away some more, in order to reduce
> himself to a state of nondoing [*wu-wei*].
> And when he has reached a state of nondoing, there is also no
> longer anything that he consciously values.

Laozi sometimes describes this end state as "naturalness" or "so-of-itself" (*ziran,* 自然). Like the plain, uncarved block, the fully natural person has stripped away the gaudy paint of socialization and returned to something like his true nature, simple and pure.

The rub is *how,* precisely, to do this. Anyone able to read the *Laozi* is already infected with culture and part of the literate elite—from the same social class as the rude recluses who make fun of Confucius in the *Analects.* How do you take a pedantic scholar, corrupted by artificial tastes, and turn him back into an uncarved block? One sometimes gets the impression that Laozi thinks that just reading his text will do the trick. We see hints of a technique that will become more prominent in the *Zhuangzi* and will then be transformed into Chan/Zen Buddhist *koan* practice: the use of paradoxical language or evocative poetry to break the reader out of a conceptual rut. "The Way that can be spoken of is not the enduring Way," Laozi famously declares. "The name that can be named is not the enduring name." This paradox is dubbed a "mystery upon mystery, the door to a multitude of secrets." We are being presented with a riddle that appears to have no answer, at least according to ordinary reasoning. The same is true of well-known Zen *koans* such as "What was your face before you were born?" or "What is the sound of one hand clapping?" These paradoxes are viewed in Zen as "gates" or "doorways" to wisdom, because confronting them helps you to break out of your ordinary patterns of thinking. In the same way, we can imagine that the "mystery upon mystery" that Laozi asks his reader to contemplate is intended as a secret door to insights beyond words.

Zen was never satisfied with *koan* contemplation alone, however; it was always combined with some sort of meditative practice. Similarly, at several points in the *Laozi* we see hints that Laozi has something more concrete in mind for his reader, a set of physical meditation practices or spiritual exercises. One section, for instance, consists of a series of rhetorical questions:

Carrying on your back your encumbered earth soul,
Can you embrace the One and not let it go?
Concentrating your breath until it is supple,
Can you be like an infant?
Polishing and cleaning your mysterious mirror,
Can you leave it without blemish?

. . . .

Opening and closing the gates of Heaven,
Can you play the role of the female?

This is an exceedingly opaque passage. What in the world is an "encumbered earth soul"? What is your "mysterious mirror," and how would you go about polishing it? Where are the gates of Heaven, and how are they related to playing the role of the female? (This is starting to sound a bit pornographic, actually.) The short answer is that nobody really knows what Laozi is advocating in this passage, but the huge amount of commentary that has grown up around it, combined with the way it was later used by practicing Daoist communities, suggests that these are references to particular forms of meditation practices, breathing techniques, and perhaps sexual gymnastics. By the early centuries of the Common Era we certainly see self-proclaimed Daoists—citing Laozi as their inspiration—engaged in all of these, and typically throwing in some hallucinogenic mushrooms to boot.

Whatever their specific nature, these conceptual riddles and physical practices have a specific goal: shutting down the conscious mind in order to allow the spontaneous, "hot" system to run without interference. The state Laozi wants to foster probably looks something like what, in exercise science, is referred to as the "runner's high." The neuroscientist Arne Dietrich has done interesting work on what he calls "transient hypofrontality," the downregulation of the cognitive control regions in the prefrontal cortex that occurs during intense physical exercise. Vigorous physical activity puts enormous stress on the human body, and the body responds by temporarily shutting down parts deemed inessential, like our energy-hungry PFC. The cognitive state that results looks very much like the relaxed, natural *wu-wei* mind-set of the Laozian sage. As Dietrich observes,

> Some of the phenomenologically unique features of this state such as experiences of timelessness, living in the here and now, reduced awareness of one's surroundings, peacefulness (being less analytical), and floating (diminished working memory and attentional capacities), are consistent with a state of frontal hypofunction. Even abstruse feelings such as the unity with the self and/or nature might be more explicable, considering that the prefrontal cortex is the very structure that provides us with the ability to segregate, differentiate, and analyze the environment.

Interestingly, it appears that we get the same neurological pattern with the use of alcohol and various drugs, or a really good bout of sex, whether you prefer to play the role of the female or not. All of these activities temporarily shut down our conscious mind—our sense of awareness, orientation, and control—leaving the unconscious system in charge. When Laozi speaks of returning to the "mind of an infant," this is, from a scientific perspective, more than just a metaphor. Laozian meditation aims to disable precisely the regions of the mind that are developed, or overdeveloped, in adulthood, allowing our more basic and ancient systems to take over.

The result is an easy oneness with things, a state of going along with whatever presents itself, with no expectations and no calculation. Such perfect relaxation brings with it incredible efficacy in the world as well as social success, as we would expect from the connection between *wu-wei* and *de*. The state of true *de*—the "highest Virtue, that doesn't think itself virtuous"—represents a perfect harmony with Heaven and the Way, which gives the Laozian sage remarkable powers over man, woman, and beast. Because he thinks nothing of himself, he is valued by others; because he wants nothing, everything is given to him.

> Demanding nothing in return for his kindness, the sage
>   eventually obtains everything;
> The sage does not accumulate things,
> Yet the more he gives to others, the more he has himself.
> Having given to others, he is richer still.

The state of keeping one's conscious mind in check and being guided only by motor programs and lower-level cognitive systems is described at one point as "holding to oneness." This state of *wu-wei* not only results in personal success but also has a ripple effect on those around the sage. Ideally, the Laozian sage succeeds, through the power of his *de*, in transforming the entire world around him:

> The crooked will be whole;
> The bent will be straight;
> The empty will be full;
> The exhausted will be renewed;
> Those with less will win out;
> Those with too much will be thrown into confusion.
> Therefore the sage holds to oneness
> And in this way serves as the shepherd of the world.
> He has no regard for himself, and so is illustrious;
> He does not show himself, and so is bright;
> He does not brag, and so is given credit;
> He does not boast, and so his name endures.

As in the *Analects*, then, there is a political dimension to this vision. The *de* of the Laozian sage is so powerful that it is able to draw others into its orbit. The naturalness of the sage causes others around him to become natural as well. The end goal is a world free of artifice, hypocrisy, and excess desire, where humans live in simple harmony with each other and the natural world and there is no need to talk about morality because kindness and goodness flow forth from people spontaneously, without the slightest effort or thought.

HOW CAN YOU DESIRE NOT TO DESIRE?

This is a beautiful vision, which goes a long way toward explaining the enormous popularity of the *Laozi* throughout history. It is the most translated book in the world after the Bible and has had a considerable impact on Western culture, especially over the past few decades. Its suspicion of book learning and established social norms, along with its celebration of a radically simplified lifestyle—close to the earth and far from the unnatural distractions of city life—made it one of the inspirations of the 1960s counterculture movement. I would hazard the claim that more joints have been rolled, and more incense burned, in the presence of this book than any other in the history of mankind. And it's hard to argue with the celebration of naturalness, or suspicion of hypocrisy, that motivates the text. Who doesn't want to be in harmony with the cosmos?

It's undeniably the case, however, that a large proportion of 1960s and 1970s hippies eventually traded in their commune lifestyles for computer start-ups, and their hemp clothes for Banana Republic. Like most Western counterculture movements—from German Romanticism in the nineteenth century to many present-day cultural anthropology departments—the *Laozi* sees its Utopian ideal as a return to a Golden Age, when primitive people lived in perfect unselfconscious harmony with nature. We can see something similar in contemporary work on spontaneity. For instance, in a collection of essays exploring flow entitled *Optimal Experience*, Mihaly and Isabella Csikszentmihalyi observe, "It is reasonable to suppose that in a primitive culture that happened to be well adapted to its environment, people would be in flow most of the time, provided they were unaware of alternative lifestyles and possibilities. In such ideal-typical communities ... life choices are self-evident, doubts and unfulfilled desires are few and transitory." They cite an anthropological study of First Nations peoples in the Pacific Northwest as their example of a perfectly spontaneous, unselfconscious lifestyle, very much in the tradition of the "Noble Savage" portrayals of small-scale or exotic societies that have a venerable pedigree in Western thought.

The main problem with the Noble Savage myth is that it's precisely that: a myth. Although propagated to this day for a variety of reasons, it is demonstrably inaccurate as an account of small-scale social life, contemporary or ancient. Supposedly peaceful and harmonious indigenous cultures were, in fact, typically characterized by constant and brutal warfare (including widespread genocide and enslavement of enemies), as well as extreme social stratification. It's possible that preagricultural people enjoyed more leisure time, better digestion, and superior posture and eyesight than us domesticated, wheat-chomping moderns, but their lives were also characterized by levels of hunger, disease, cruelty, and violence that we would find shocking. Why that might be—and why, for the most part, counterculture movements tend to fade away, leaving consumerist capitalism unscathed—also tells us something about the limitations of the Laozian vision of *wu-wei*.

The best way to get our heads around the shortcomings of Laozi's strategy is to examine two tensions that lurk behind the mystical poetry of the text. The first has to do with the picture of human nature presented by the *Laozi*. The author(s) of the text argue that their society, to "return" to naturalness, needs to be radically downsized:

Reduce the size of the state and decrease its population.
See to it that labor-saving devices are not employed,
See to it that people have a proper fear of change and do not move
    to distant places,

See to it that,
    although they have carts and boats, the people do not ride in
        them;
    although they have armor and weapons, they do not use
        them.
See to it that the people return to using the knotted rope,
That they
    find sweetness in their food
    beauty in their clothes
    joy in the customs
    contentment in their homes.
See to it that, though neighboring states are within sight of one
    another
and the sounds of chickens and dogs can be heard across the
    border,
the people will grow old and die without ever having traveled
    abroad.

This sounds eerily like the hypothetical "primitive culture" described by the Csikszentmihalyis. The "knotted rope" in this passage refers to a method of record keeping said to have been deployed before the advent of writing, which gives you a sense of how far back Laozi wants us to go. We have to give up not only ox-drawn plows but also literacy itself. This is somewhat awkward in the sense that it's a written text telling us that writing is wrong, but presumably the *Laozi* is intended as a message that self-destructs upon delivery. You read it, your mind is erased, and you then spend the rest of your days blissfully embracing the uncarved block.

But we're still left with the question: If it's so natural to be like the uncarved block, if all we're being asked to do is return home, why do we need a book to tell us to do it? The fact that we *do* need it suggests that maybe being simple is not so instinctive after all. The fact that the Laozian ruler is going to have to "see to it" that the common people don't use labor-saving devices or travel abroad suggests that they are naturally inclined to do precisely both of these things, and much worse as well.

This tension might be dismissed as merely inaccurate anthropology or psychology. It is, however, linked to the related, but much deeper, problem of how you try not to try. The *Laozi* is full of warnings about the dangers of trying. It warns, for instance, that anyone who tries to have an effect on the world is doomed to failure:

> I declare that those who wish to take the world and do something
> to it will not be successful.
> The world is a sacred vessel—not a thing that can be worked
> upon.
> Work on it and you will ruin it;
> Try to grasp it and you will ruin it.

At the same time, the text also urges us to "do *wu-wei*," or to "grasp the image" of the Way so that we can control the world and escape from the cycle of reversion. Moreover, it promises that if we stop trying, we'll in fact be very successful. The sage "does nothing, and yet nothing remains undone."

There are, when we look more closely, many other points in the text where we appear to be presented with crude instrumental reasoning disguised as mystical wisdom, passages that seem primarily concerned with helping us to get ahead in the world. "The rivers and seas rule over the hundred valleys because they are good at putting themselves below them," begins one of these passages. "By putting themselves below, they end up on top." The reference is to a common trope in early Chinese texts that has the rivers and seas, the lowest places in the world, "ruling over" the valleys by gathering all of their water to them. The text then spells out the lesson to be learned from this for the human world:

> In the same way, if you desire to rule over people,
> you must, with your words, put yourself below them.

If you want to be in front of the people,
> you must, in your actions, put yourself behind them.

This is why the sage is able to take a superior position but the
> people do not resent it,

able to take the lead without the people minding.

It is why everyone in the world is happy to help him along and
> never tires of it. Because he does not contend with others,

there is no one in the world who is able to contend with him.

It is in passages like this that the "instrumental" strand of the *Laozi*—advice that seems practically, and somewhat sinisterly, designed to help rulers and others get ahead in the world—becomes most salient. It's no coincidence that this is also where the paradox of *wu-wei* becomes most evident. What we see here seems to be advice about how to *fake wu-wei*: to appear to value weakness in order to gain strength, to appear to be humble so that people won't mind when you stick a crown on your own head.

Interestingly, this instrumental problem is something that the Csikszentmihalyis note as a tension that arises in advocating "flow": "Most people are unimpressed by the fact that flow provides an optimal subjective experience, but their interest immediately perks up at any suggestion that it might improve performance. If it could be demonstrated that a fullback played harder if he was in flow, or that an engineer turned in a better product if he was in flow, then they would immediately embrace the concept and make a great deal of it. This, of course, would effectively destroy the autotelic nature of the experience." They conclude that it is therefore "probably better to downplay" the connection between flow and enhanced performance.

That may be an option for them, but it's certainly not for Laozi, since the very premise of his thought is that valuing the weak and the lowly will allow you to become, in the end, strong and prestigious. The main selling point of the text is that the sort of *wu-wei* that it advocates *works*. In fact, the *Laozi* has often been used for blatantly instrumental purposes. Laozian teachings about the strength of weakness inform Sunzi's *Art of War*, which teaches how to feign vulnerability—to tactically retreat—in order to spring the perfect ambush and wipe out your enemy. The so-called "soft" martial arts are based on a similar principle: give way, be "weak," so that you can use the momentum of your opponent against him and whack him on the back of the head as he stumbles by.

The instrumental reading of Laozian *wu-wei* is, arguably, not in keeping with the original spirit of the text. Laozian *wu-wei* works only if you're sincere: you can attain power only if, at some level, you truly don't want it. Again, as with the *Analects*, we need to transform our basic *values*, not just our behavior. Chapter 20 of the *Laozi* presents a wonderful portrait of what it would be like to live without an active cognitive control system, guided solely by more basic processes, and it links the ability to maintain this state to holding fast to certain values:

> Most people are loud and boisterous,
> > as if attending a lavish feast or off on holiday.
> I am instead tranquil and make no display,
> Like an infant that has not yet learned to smile,
> Drifting as if with no home to return to.
> Most people possess more than they need.
> I alone seem in want.
> I have the mind of a fool—simple and blank.
> Most people are bright and clever,
> I alone am muddled.
> Vast, like the ocean.
> Endless, as if never stopping.
> Most people have conscious goals and lofty purposes.
> I alone am ignorant and uncultured.
> My desires alone are different from those of others
> Because I value being fed by the Mother.

This focus on embracing a particular type of value, however, just makes the tension involved in attaining *wu-wei* even more profound. In order for you to be genuinely *wu-wei*, your basic desires need to change. You need to come to genuinely value "being fed by the Mother"—clinging to weakness, darkness, and ignorance, and thereby being truly in touch with the natural Way. You need to "desire not to desire."

So it seems that we have the same basic problem we saw in the *Analects*: we are being asked to love something we don't already love, to value something we don't already value. The only way you can get the benefits of Laozian *wu-wei* is if you sincerely don't want them. How could that possibly work?

One thinker in particular saw this problem clearly, a follower of Confucius named Mencius. Arguing that Laozi's primitivism is flawed from the very beginning, Mencius focuses on the Laozian picture of human nature. The fact that people like labor-saving devices, luxuries, cleverness, and traveling abroad means that there is something natural about these things, at least when pursued within reasonable bounds. He also thinks he's found a way around the paradox that emerges in both the *Analects* and the *Laozi*: attaining *wu-wei* does not require coming to love something we don't already love, or sincerely embracing a completely new set of values. According to Mencius, we are, at some level, *already wu-wei*. We just need to realize it and thereby become capable of nourishing the *wu-wei* tendencies within us.

5

Try, but Not Too Hard

CULTIVATING THE MORAL SPROUTS

IN WARRING STATES CHINA, THE REMNANTS OF THE DEFEATED Shang royal line were allowed to take refuge in the state of Song. Perhaps because of its association with the defeated dynasty, the "man from Song"—humorously dim-witted and inept—became the standard butt of jokes throughout later Warring States texts. Probably the first time he pops up is in the *Mencius:* "In the state of Song there was a man who, worried because his sprouts of grain were not growing fast enough, decided to go out to his field and pull on them. Without any idea of what he'd done, he returned home and announced to his family, 'I am really exhausted today, I've been out in the fields helping the sprouts to grow!' Alarmed, his sons rushed out to the fields to take a look and saw that all the sprouts had shriveled and died." This episode is the basis of a modern Chinese saying, "pulling on the sprouts to help them grow," which refers to any effort that has thoroughly counterproductive results. Mencius intends this to be a message about the proper way to attain morally correct *wuwei:* "You must put some work into it, but you can't force it. Do not be like the man from Song!" It's clear that he has two separate targets, those who try too hard (the man from Song) and those who don't try at all: "There are those who think that there is nothing they can do to help their moral growth, and therefore abandon all effort

entirely. They are the people who fail to weed their sprouts. Then
there are those who try to help their sprouts to grow, the sprout-
pullers. Not only do their efforts fail to help, they do active harm."

Mencius was a self-professed follower of Confucius and saw his
job as defending Confucianism against a variety of philosophical
threats, including the two that are being singled out in the Farmer
of Song story. First we have the Laozian primitivists we met in the
last chapter, whose advocacy of *wu-wei* in the more literal sense of
"doing nothing"—rejecting culture and self-cultivation—means
they are "failing to weed their sprouts." The second threat comes
from the "sprout-pullers," a school called the Mohists, who rejected
wu-wei entirely and advocated an entirely rationalist model of ethics
that wouldn't look out of place in the modern West. Mencius's agricultural model for moral cultivation, which harmonizes the need
for effort with the ideals of naturalness, can be seen as a response to
both schools. At the same time, it also manages to solve problems
with both the Confucian and Laozian strategies for attaining *wu-
wei*.

Although he loudly professed his commitment to Confucianism,
Mencius was so deeply influenced by Laozi's celebration of "naturalness" that he modified Confucius's thought in fundamental ways,
which is why we'll be discussing him on his own, as opposed to lumping him together with Confucius and Xunzi. Dedicated to Confucian values, but drawn to the idea that *wu-wei* should spring from
our innate tendencies rather than being forced upon us from the
outside, Mencius formulated a unique version of Confucianism that
is still very appealing today. Mencian thought attempted to chart
a course between the traditionalism of Confucius and Xunzi and
the radical primitivism of Laozi. Confronted with opponents who
argued that the answer to the social chaos of the Warring States was
for everyone to adopt a strict, rationalist code of morality, he argued
that the Confucian emphasis on *wu-wei* was instead the right way
to go. We do indeed need to foster morality, but any viable model

of moral education has to be grounded in embodied spontaneity. When it comes to being moral, you need to try, but not too hard, and in a way that does not go against your natural tendencies. You certainly don't want to be yanking on your sprouts.

AGAINST THE RATIONALISTS: COLD CAN'T GO IT ALONE

The Mohists, members of a school founded by Mozi 墨子 ("Master Mo"), shared Confucius's concern about the moral chaos of their age but were appalled by the Confucian emphasis on emotional training, immersion in traditional cultural forms, and role-specific social duties. To their minds, this strategy was a recipe for fuzzy-minded thinking, nepotism, and cronyism, not to mention an enormous amount of wasted time and money. They advocated a radical reorganization of society along completely rational lines, with the guiding principle being the maximization of the material wealth, population, and order of the state. The elaborate rituals and costly musical performances at the heart of Confucian practice would be banned: they kept healthy young people out of the fields and workshops, where they could actually do something useful. Filial piety would be completely reimagined, directed away from one's own biological parents and toward *all* the parents in the state. No favoritism allowed.

Mohist moral "training," to the extent we can use that word, was a matter of learning one simple principle: how to calculate the consequences of any given act and then determine if those consequences would, overall, increase three things: the wealth, population, and order of the state. If the answer was no, the act was wrong. Moral decision making was thus radically simplified. No more lifetime cultivation of refined moral sensibilities, to be applied in a context-sensitive, flexible manner. This Confucian approach, in Mozi's view, was elitist and ripe for abuse. Moral decision making should be objective, like doing a simple equation. Weigh up the positive and negative consequences, do the math, and then force yourself to go along with whatever course of action the results dictate.

For instance, let's say your father dies. Your immediate emotional response is to honor him with extended mourning and send him off with an elaborate burial. Would withdrawing from work and spending a huge amount of money on a coffin, burial plot, and associated paraphernalia increase the wealth, population, and order of the state? Absolutely not: your work productivity would dive, and material resources would be needlessly wasted. Therefore, bury your father in a simple, unmarked grave and get back to work. This will go against your emotional inclinations, but that's because your emotions are chaotic and selfish. Social order requires that they be overridden by the rational mind.

Brutal, perhaps, but this model of ethics—nowadays referred to as "utilitarianism" or "consequentialism"—is very much alive and well, and is in fact one of the dominant modes of ethical reasoning in the modern West. Modern variants look a little less weird because they typically aim to maximize things (like happiness or freedom from suffering) that are intuitively more appealing to us than Mozi's criteria (the wealth, population, and order of the state). Modern utilitarianism is still perfectly capable, however, of producing demands that, though perhaps rationally defensible, sit uneasily alongside what we might consider normal emotional inclinations. For instance, the philosopher Peter Singer, perhaps the best-known contemporary utilitarian, has concluded, among other things, that euthanasia for severely disabled infants might, in some instances, be justifiable and that, as long as global poverty remains anywhere near current levels, it is immoral not to give away all of one's income above the amount needed to maintain a minimum level of material well-being.

Singer's conclusions strike many, especially nonphilosophers, as a bit out of tune with normal human psychology. While we can concede that, rationally, it is hard to justify buying our own five-year-old a bicycle when there are millions of five-year-olds around the world without enough to eat, people all over the industrialized world continue to buy their children bicycles. Moreover, while we can perhaps admire saints like Mother Teresa, who give away everything to serve others, most of us admire them from a distance. There's something slightly inhuman about extremes of altruism, and we demand such extreme behavior neither of ourselves nor of our friends.

To the argument that pure rational utilitarianism goes against normal human psychology, Singer's reply is, so much the worse for human psychology. Human nature is something to be overcome, not indulged, and Mozi had the same response to those who argued that his teachings were unrealistic. Mencius's counterargument was that psychological realism is crucial. If your model of ethics asks people to do something that they cannot, in reality, do, that's a serious problem. To begin with, we have the moral saint problem. If I tell you that being moral requires giving away most of your salary, or quitting your job and dedicating your life to serving the poor, the likely result is that you're going to simply give up on morality altogether. Pulling too hard on the sprouts kills them; they shrivel up and die. Mencius was adamant that if moral goodness was your goal, you needed to give people a step-by-step, attainable path to reach it.

A related issue is that human beings naturally care more about their family than their friends, and about their friends more than strangers. Any model of morality that swims against this current, Mencius argued, is doomed to failure. For Mencius, abstract thought is not a strong enough foundation to support morality—only embodied emotions have the motivational power, speed, and flexibility to guide proper behavior in the real world. The goal of education should not be to teach people logic and self-control but rather to guide them in nurturing a set of positive, innate tendencies into full *wu-wei* dispositions.

Extending the agricultural metaphor, Mencius called these

tendencies the "four sprouts." We can get a sense of what these sprouts feel like psychologically by invoking Mencius's "child and the well" thought experiment:

> Now, imagine a person turning around and, all of a sudden, spotting a small child stumbling toward the opening of a deep well. There is no one who, in such a moment, would not experience a feeling of alarm and empathy. Their response would be motivated by this feeling alone—not because they want to save the child and thereby gain some merit with the parents, not because they want to gain a reputation for goodness among their neighbors and friends, and not because they want to avoid having to hear the child's anguished cries. From this we can see that someone lacking this feeling of empathy cannot be called a proper human being.

This "feeling of empathy," Mencius explains, is the sprout that, if properly cultivated, will grow into the full virtue of benevolence or compassion. Another sprout is that of "righteous indignation," the beginning of moral correctness: "Imagine that one is so hungry that a single plate of rice, a single bowl of soup, is the difference between life and death. If such food were served with insults and abuse, even a homeless person would not accept it; if it were handed out after having been trampled upon and then scraped off the ground, even a beggar would refuse to eat it." Mencius is claiming that, no matter how humble our circumstances, there are things we simply won't do, and this reaction is driven by a kind of righteous indignation that will cause us to reject offers—even life-saving offers—that we consider beneath our dignity. These sprouts of compassion and moral correctness are rounded out by the "feeling of deference" (the sprout of ritual propriety) and the "feeling of right and wrong" (the sprout of wisdom). These four virtues were, by Mencius's time, the four primary Confucian virtues, and they stand in for all the virtues in general. According to Mencius, all humans have them just as we all have "four limbs," and they are implanted in us by Heaven. To develop them is to serve Heaven; to neglect them is to waste our Heavenly endowment.

Bracketing Mencius's theological language, we can, from a scientific perspective, find quite a bit of support for some of his basic claims. To begin with, there is considerable evidence from evolutionary biology and cognitive science for the existence of discrete, innate moral emotions, not only in humans but also in many mammals. Most of this work has focused on empathy—literally, "into-others + feeling"—which is a basic mammalian instinct underlying a wide variety of social interactions. In humans and other primates, it seems to involve the so-called "mirror neuron" system, which creates a kind of sensorimotor resonance when we observe, or even imagine, actions involving other agents. When you see someone cut himself and find yourself flinching, that's the mirror neuron system in action. The current dominant theory is that empathy, enabled by the mirror neuron system, is what motivates compassionate behavior.

Mencius had no access to fMRI technology, but his "child and the well" scenario captures this process beautifully. When we see a child about to fall into a dark, cold, scary well, our mirror neurons fire, allowing us to experience the same alarm the child is about to feel, which in turn immediately triggers empathy. In normal human beings, this leads directly to compassionate behavior: we run to save the child without thinking or deliberating. This research also explains breakdowns of compassion. There is considerable evidence that human psychopathy involves deficits in mirror-neuron and emotional regions of the brain. Similarly, attempts by governments or leaders to reduce compassion toward particular groups—tribal enemies, persecuted minorities—typically focus on short-circuiting it by portraying the victims as vermin or germs, and therefore not worthy of our empathy.

Similarly, the sprout of "righteous indignation" finds a parallel in contemporary science and serves as further indication that, in human social interactions, rationality is not in the driver's seat.

Imagine an economic game, called the Ultimatum Game, where you are given $100 to split between yourself and another person. You have to give some of it away but get to decide the actual amount. The catch is that the other person can veto the whole transaction. If he or she says no, neither of you gets any money. There is only one rational strategy, from a purely cold-cognition perspective: given $100 to split, you should offer $0.01, keeping $99.99 for yourself, and the other person should accept, since both of you end up better off financially than before. In reality, no one makes such low offers (would you?), and people on the receiving end will indignantly reject any deal that seem grossly unfair, the usual rejection threshold hovering around 20 to 30 percent. Neuroimaging studies of subjects playing the Ultimatum Game show that powerful emotion drives the rejection of unfair offers; other studies suggest that behavior in this game may have a genetic basis, an indication that righteous indignation might be part of our evolved psychological architecture.

Like compassion, this trait may not even be limited to humans. One study found that capuchin monkeys refuse food—out of moral outrage, we are tempted to say—when they see another monkey getting more cucumbers than it deserves. Mencius attributes perhaps too much strength to this feeling, in the sense that it's not at all clear that most people would choose death by starvation over dishonor. The idea that humans might have an innate "sprout" of righteousness, however, seems plausible. How the other two sprouts (the beginnings of wisdom and ritual propriety) would map onto other aspects of our moral psychology is less obvious. Humans clearly do, though, possess other, distinct moral emotions, such as conceptions of purity derived from disgust reactions, that are promising candidates.

More generally, the Mencian view that morality is about emotion-driven, *wu-wei* behavior is becoming the dominant view among psychologists. We've already discussed the broader finding that unconscious, "hot" processes (emotions, habits, implicit skills) typically play a much greater role in determining our behavior than conscious, "cold" processes. A parallel development in philosophy is the growing recognition—at least among those philosophers inclined to look beyond their armchairs—that moral judgments are driven by emotions to a much greater extent than we tend to realize. A group of young, psychologically informed philosophers dubbed the "neo-Humeans" have been reviving the work of David Hume (1711–76), an Enlightenment thinker who argued, very much in the minority, for giving emotions their due. In psychology, researchers like Jonathan Haidt have been demonstrating that people's moral judgments are the result of hot, fast, visceral reactions to social scenarios. When questioned, people will often invent plausible-sounding rationales for their decisions, but clever experimental manipulation shows that these typically come after the fact—the "rational tail" wagged by the "emotional dog."

Even the idea that we can cleanly separate rationality from emotion is coming under fire. Particularly relevant in this regard is the groundbreaking work of the neuroscientist Antonio Damasio on patients suffering from damage to the ventromedial prefrontal cortex (VMPFC). The VMPFC is a center of emotion processing in the brain, particularly complex, learned emotions. The patients studied by Damasio had suffered accidents or strokes that caused very localized damage to this region, sparing their rational faculties—their hot cognition was impaired, but not their cold. Despite their difficulty processing emotion, they had no trouble when it came to memory, abstract reasoning, math, or standard IQ tests. Nonetheless, they were barely capable of functioning. When it came to real-life decisions, they were incredibly inept, apparently incapable of making simple choices or taking into account the future consequences of their actions.

Damasio and his colleagues have come to believe that these patients lack "somatic markers"—the unconscious assignments of emotional value that ordinarily accompany our representations of the world. According to Damasio, our images or mental pictures of the world—whether of people, places, or things—are imbued with feelings of goodness or badness, urgency or lack of urgency, by our hot emotional systems, and these feelings then play a crucial role in anything we'd consider normal decision making. This is because, in any given situation, the number of theoretically possible courses of action is effectively infinite. Couple this with the fact that the conscious mind has limited capacity, and we've got a problem. Left alone, without the help of our hot cognition, cold cognition is simply paralyzed by choice. Therefore, the hot system normally helps out by biasing the reasoning process, usually unconsciously, with somatic markers before it even begins. Patients with damaged VMPFCs, however, are deprived of this guidance, forced to drift along in a world where every option feels as good or bad as any other option.

Revealingly, they continue to function perfectly as rationalist philosophers. Given an artificially simplified moral dilemma, they can do the math or follow the rules properly. Real life, though, is much more complicated. Deprived of the filtering function of somatic markers, these patients become paralyzed by indecision or simply commit themselves to randomly selected courses of action. In a way, they serve as a controlled experiment demonstrating that rationalist models of ethics are radically incomplete: these people have everything that Mozi or Peter Singer says a moral person needs—a perfectly functioning cold-cognition system—and yet they are a far cry from what we'd consider competent moral agents.

The conscious mind, ungrounded by the wisdom of the body, is remarkably incapable of taking care of business. Damasio labels as "Descartes' error" the idea that morality, or any human behavior, can be guided solely by disembodied reason. Mozi made the same mistake. Morality in the real world has to be spontaneous, unselfconscious, automatic, and *hot*. Coming to some rational conclusion about the right way to act, and then trying to force your body to comply, simply doesn't work. It's not effective at the individual level, and it's not sustainable on the social level. Realistic moral behavior has to spring from the spontaneous, embodied mind—from *wu-wei*.

This does not necessarily eliminate the need for instruction and training. As we've discussed, the unconscious can be educated. We're not born knowing how to ride a bike or drive a car. In the same way, although we appear to be born with some nascent moral feelings, the behavior of your average three-year-old suggests that additional training and practice are required to become a mature person. This is the essence of Mencius's critique: you can't try to consciously *force* morality, like the sprout-pulling Mohists, but you also can't just sit around, listening to a Jerry Garcia solo and hoping socially desirable behavior will somehow magically follow.

AGAINST THE PRIMITIVISTS: CULTIVATING YOUR MORAL GARDEN

At one point in the text that bears his name, we find Mencius in the small state of Teng, whose king has established a kind of non-partisan think tank where scholars from many different schools have gathered. Among them are a group of Primitivists who worship a legendary figure known as the "Divine Farmer." They seem inspired by Laozi-like teachings, being committed to a rural lifestyle, simple technology, a rudimentary economic system, and complete social equality—arguing, for instance, that rulers should till the land beside their "subjects." One of them approaches Mencius to make his case and runs into a wall of philosophical scorn. Mencius exposes their rather tenuous grasp of basic economics and argues that it is only the advent of Confucian culture that managed to pull human beings out of a brutish, uncomfortable existence. He concludes, "I have heard of coming out of the dark valleys to settle in the tall trees, but not of abandoning the tall trees to descend back into the dark valleys"—it would be perverse to give up the benefits of cultivated Confucian society in order to return to the benighted, primitive existence foolishly celebrated by the *Laozi*.

A dominant Laozian metaphor for naturalness is the uncarved

block, a simple, unadorned, unworked chunk of wood. Mencius replaces this static metaphor with a dynamic one: nature is like a sprout that wants to grow in a certain direction, it just needs a little help. The primitive stasis that Laozi celebrates does not represent true naturalness but rather a stunting of natural human tendencies. Mencius was dedicated to naturalness in a *cultivated* sense, not the wild, weedy state of primordial nature. For him, as for Confucius, any *wu-wei* worth having was an achievement and required training and effort. The difference is that Mencius believed that we are blessed with the beginnings of morally proper *wu-wei* within us—the four sprouts. Although we are not spontaneously good when born, we *tend* toward good, in the same way that wheat sprouts will grow into mature plants if given the proper environment and care. Becoming spontaneously good involves merely developing these tendencies, under the guidance of a wise teacher, who plays the same role as a knowledgeable farmer. You need to apply fertilizer at the right times, weed when necessary, and ensure proper irrigation.

Consider the child-and-the-well story recounted above. By presenting us with this thought experiment, Mencius has already begun the cultivation process. He invites us to imagine this scene—an innocent toddler about to fall to his death—and then to analyze our emotional reaction. Would we care about our moral reputation? No. Would we worry about our ears being offended by the dying child's cries? No. What would we feel? Pure sympathetic alarm and empathy. This is the first task of the *wu-wei* farmer: identify the sprouts of goodness, and learn to distinguish them from the weeds. Once the sprout is isolated and identified, the next step is to help it to grow. In other words, having identified the stirrings of the *wu-wei* that we would like to attain, we need to focus on it, strengthen it, and extend it.

How precisely this might work is illustrated in a famous exchange between Mencius and a ruler he is trying to reform, King Xuan of Qi. King Xuan is a corrupt, brutal tyrant. His people live in a constant state of terror, suffering under the weight of heavy taxes, an arbitrary legal system, and capricious military drafts and labor levies. Mencius is visiting the court in an attempt to improve the king's behavior and has been lecturing him about the duties of a *true* Confucian king. Both recognize that, despite his title, Xuan falls short of this ideal. Mencius has had little success, and King Xuan finally protests that he is simply incapable of being good. He couldn't care less about the welfare of the common people. His *wu-wei* tendencies are purely bad, and there's nothing he can do to change that. He enjoys attaining glory in aggressive wars against his neighbors, hunting in his game park, and kicking back in the palace with some wine and women. That is where his spontaneous inclinations lead him.

Mencius responds by recounting an anecdote that he heard from one of the king's ministers. One day, the minister reported, an ox was being led to slaughter so that its blood could be used to consecrate a new bell. The king supposedly declared that he could not bear the animal's terrified appearance—eyes rolled up into its head, bellowing in fear—and ordered that it be spared, with a lamb substituted in its place.

> "I wonder," Mencius asked, "is there any truth to this report?"
>
> "There is," the king replied.
>
> "That feeling that you had when you saw the ox—this feeling alone is sufficient to enable you to become a true king. The common people all thought that you begrudged the additional expense of the ox, but I knew for certain that it was because you could not bear to see its suffering."
>
> "That's true," the king said. "The common people may talk, but even though Qi is a small state, we can certainly afford the expense of a single ox! It was simply that I could not bear its look of abject terror, like an innocent man going to the execution ground, that I substituted a lamb in its place."

Mencius then points out that the common people cannot be blamed for thinking the king miserly: if avoiding animal suffering was his concern, what difference does it make whether the slaughtered animal is an ox or a lamb? The king concedes that Mencius is right, and expresses puzzlement over his own actions. "What was I thinking?" he wonders aloud.

> "There is nothing puzzling about this," Mencius declared. "The motive behind your action is the very means by which one attains compassion. You saw the ox; you never saw the lamb. The attitude of the gentleman toward animals is this: having seen them alive, he cannot bear to see them die; having heard their cries, he cannot bear to eat their flesh. That is why the gentleman keeps his distance from the kitchen."
>
> Amazed, the king replied, "In the *Book of Odes* we read, 'It is the other person who has the feeling / But it is I who can gauge it.'
>
> "This describes you, Mencius. Even though the action was mine, when I looked into myself for a motivation I could not discern my own true feeling. Your words have gone straight to my heart and profoundly moved me."

Here we see Mencius in the role of moral psychoanalyst. Through a combination of psychological insight and careful questioning, he has enabled the king to grasp his own true motivation, which had hitherto been opaque—not only to his subjects but also to the king himself. In doing so, he forces the king to admit that he possesses the "sprout of empathy," something that the king himself has denied.

Having conceded this point, the king still resists. What does a fleeting moment of compassion for a terrified animal have to do with being a true king? Mencius now moves in for the rhetorical kill.

> "If someone said to you, 'I have enough strength to lift a hundred pounds, but not enough to lift a single feather; my eyes can perceive the tip of a fine hair, but cannot see a huge cart of wood,' would you find this reasonable?"

"No, I would not."

"As for you, King, your kindness is sufficient to reach an animal, and yet the benefits of your rule fail to reach the common people. Are you any different from this hypothetical person? That a single feather is not lifted is because no effort is applied; the cart is not perceived because the eyes are simply not being used. That the common people are not cared for is because your kindness is not directed toward them. Therefore, your failure to become a true king is due to a refusal to act, not an inability to act."

His compassion toward the ox proves that the king has the emotional capacity to be a true king. All he needs to do is *exercise* this capacity, so that it can reach its proper beneficiary: the common people. The way to make this happen is through a combination of introspection, practice, and imaginative extension.

The king needs to reflect on the way he felt when he saw the terrified ox—ruminate on it, roll it over on his moral tongue, as it were—to fully appreciate what empathy in the presence of a suffering, sentient being feels like. Then, as Mencius puts it later in the conversation, he needs to work on "taking this feeling here" (sympathy for the terrified ox) and applying it to "what is over there" (empathy for his terrified, oppressed people). Mencius also suggests trying to expand the circle of concern by beginning with familial feelings. Focus on the respect you have for the elders in your family, he tells the king, and the desire you have to protect and care for your children. Strengthen these feelings by both reflecting on them and putting them into practice. Compassion starts at home. Then, once you're good at this, try expanding this feeling to the old and young people in *other* families. We have to imagine the king is meant to start with the families of his closest peers, who are presumably easier to empathize with, and then work his way out to more and more distant people, until he finally finds himself able to respect and care for the commoners. "One who is able to extend his kindness in

this way will be able to care for everyone in the world," Mencius concludes, "while one who cannot will find himself unable to care for
even his own wife and children. That in which the ancients greatly
surpassed others was none other than this: they were good at extending their behavior, that is all."

Mencian *wu-wei* cultivation is about *feeling* and *imagination*, not
abstract reason or rational arguments, and he gets a lot of support on
this from contemporary science. The fact that imaginative extension
is more effective than abstract reasoning when it comes to changing
people's behavior is a direct consequence of the action-based nature
of our embodied mind. There is a growing consensus, for instance,
that human thought is grounded in, and structured by, our sensorimotor experience of the world. In other words, we think in images.
This is not to say that we necessarily think in pictures. An "image"
in this sense could be the feeling of what it's like to lift a heavy object
or to slog in a pair of boots through some thick mud.

The key feature of an image is that it's *analog* rather than *digital*.
To understand the difference between the two, consider audio recording. I was introduced to recorded music through 45 rpm vinyl
records. I later moved on to full-sized records and eight-track tapes
(I still have a vivid memory of my first eight-track, Elton John's
Goodbye Yellow Brick Road, which I played until it broke). These
were *analog* technologies: Elton John played music in a studio, and
a device there allowed the resulting sound waves to impress themselves upon a physical medium in a more or less direct fashion. The
analog nature of the recording meant that it was tied to its medium
in an important way, because the only way to get the music from a
record onto, say, a cassette tape was to allow the analog patterns of
the sound waves from the vinyl record to create new analog patterns
in the magnetic ribbon of the cassette. In an important sense, an
analog recording is always directly connected to the experience that
gave rise to it—there was a direct physical link between Elton John
playing the piano somewhere and the eight-track tape that I stuck
into my nifty new stereo.

When it comes to *digital* recordings, on the other hand, this link is broken. Digital recordings take the sound waves produced by an actual person playing or singing and translate them into an entirely different medium: strings of os and 1s that have no direct connection to the original performance. It's only abstract information, not a physical impression. The disembodied model of human cognition holds that most, if not all, concepts are *digital* in precisely this way. They may have originated in some experience of the world, but have since been translated into an entirely different medium, their link with the original experience completely severed.

For many decades, cognitive scientists and philosophers have been debating whether the mind is essentially analog or digital. The debate is by no means over, but the eight-track camp has the upper hand and seems to be gaining. One of the basic problems with the disembodied theory is that our minds seem to have been built before the digital revolution. No one would deny that our sensory organs are analog: sound waves and light waves impress themselves on our embodied mind in much the same way that analog recordings or photographs are made. However, even the "higher" regions of our brain, like the cerebral cortex, look a lot more like an old eight-track player than my iPhone. The neural architecture in the cortex seems built to manipulate two-dimensional, imagistic maps rather than digital symbols. Of course, the cortex *could* contain digital processing systems. But we haven't seen them in action yet, and it's hard to imagine how and where the analog signals coming from our senses would get translated into a digital format. There is also a huge and constantly growing body of experimental evidence suggesting that the way that people actually think looks imagistic rather than abstract. We think of our lives as journeys, reason about fairness by drawing upon physical balance, and viscerally experience evil as darkness or pollution, good as light and purity.

Here again, Mencius seems prescient. The Mohists, like their modern utilitarian cousins, think that good behavior is the result of *digital* thinking. Your disembodied mind reduces the goods in the world to numerical values, does the math, and then imposes the results onto the body, which itself contributes nothing to the process. Mencius, on the contrary, is arguing that changing your behavior is an *analog* process: education needs to be holistic, drawing upon your embodied experience, your emotions and perceptions, and employing imagistic reflection and extension as its main tools. Simply telling King Xuan of Qi that he *ought* to feel compassion for the common people doesn't get you very far. It would be similarly ineffective to ask him to reason abstractly about the illogical nature of caring for an ox while neglecting real live humans who are suffering as a result of his misrule. The only way to change his behavior—to nudge his *wu-wei* tendencies in the right direction—is to lead him through some guided exercises. We are analog beings living in an analog world. We think in images, which means that both learning and teaching depend fundamentally on the power of our imagination.

In his popular work on cultivating happiness, Jonathan Haidt draws on the metaphor of a rider (the conscious mind) trying to work together with and tame an elephant (the embodied unconscious). The problem with purely rational models of moral education, he notes, is that they try to "take the rider off the elephant and train him to solve problems on his own," through classroom instruction and abstract principles. They take the digital route, and the results are predictable: "The class ends, the rider gets back on the elephant, and nothing changes at recess." True moral education needs to be analog. Haidt brings this point home by noting that, as a philosophy major in college, he was rationally convinced by Peter Singer's arguments for the moral superiority of vegetarianism. This cold conviction, however, had no impact on his actual behavior. What convinced Haidt to become a vegetarian (at least temporarily) was seeing a video of a slaughterhouse in action—his *wu-wei* tendencies could be shifted only by a powerful image, not by an irrefutable argument.

In one passage, Mencius compares the process of self-cultivation to a general's leading of his troops, with the general standing in for the mind (cold cognition) and the troops for the body (hot cognition). Without the guidance of the mind, the body would be unable to develop in the right direction, in the same way that troops without a leader would just mill about aimlessly or rush confusedly into battle. At the same time, the general on his own is fairly powerless: if he is to be effective, he needs his troops at his back and must urge them onward lest they rebel or dissolve into a confused mob. Ultimately, the general's commands can go only so far. Like a farmer patiently tending a plant, or a flood-control engineer attempting to channel the mighty Yellow River, cultivation works only when it is in harmony with the natural tendencies of things.

GIVING IN TO THE BEAT: MENCIAN *WU-WEI*

Mencius's concept of self-cultivation should recall the carving-and-polishing strategy of Confucius and Xunzi, in the sense that cold cognition is being used to change hot cognition. We can get a clear sense of how Mencius's strategy differs from theirs, however, by comparing their uses of the same metaphor, dance, to make very different points.

In a section where Xunzi is using classical dance to stand in for Confucian cultural training in general, Xunzi asks rhetorically, "How can we understand the meaning of dance?" This is his answer:

> I say the eyes by themselves cannot perceive it and the ears by themselves cannot hear it. Rather, only when the manner in which one gazes down or looks up, bends or straightens, advances or retreats, and slows down or speeds up is so ordered that every movement is proper and regulated, when the strength of muscles and bones has been so thoroughly exhausted in according with the rhythm of the

drums, bells, and orchestra that all awkward or discordant motions have been eliminated—only through such an accumulation
of effort is the meaning of dance fully realized.

For both Xunzi and Confucius, *wu-wei*—the complete absence of
effort—can be obtained only through an "accumulation of effort."
You pile effort on top of effort until what you've learned becomes
so internalized the effort falls away. The process involves hard work
and a complete transformation of the self; as with carving a piece of
jade or wood, the end product looks nothing like what you started
with. The danger with this strategy, as we noted, is that it might not
work. Back in my early thirties I took quite a few salsa lessons, and
eventually learned to follow the steps with some proficiency, but I
never really got beyond the stage of the village poseur, who just goes
through the motions. I never learned to truly like or enjoy salsa, and
others could sense it. Even after I stopped stepping on people's toes,
I was no one's first choice as a dance partner. Maybe it's because I
didn't persevere enough—this is what Confucius and Xunzi would
say—but maybe I just don't have salsa in my soul, and trying to put
it there was profoundly misguided.

Mencius, on the other hand, believes that we all have salsa in
our souls. Speaking of the pleasure one experiences when contem-
plating ancient moral exemplars and hearing morally edifying classical music, Mencius employs his own dance metaphor, declaring:
"When such joy is born, it cannot be stopped. Since it is unstoppa-
ble, you cannot help but begin to unconsciously dance along with
your feet and wave your hands in time with it." Notice that he says
nothing about training. For Mencius, when you've really learned to
focus on your true nature, your hands and feet just spontaneously
begin moving in time to the Confucian rhythm.

Of course, shutting down salsa classes—and any other type of artistic or leisure activity—would be the Mohists' first act upon taking
power. For them, the body contributes nothing to ethics, which
means that bodily strength should be expended solely on practical

tasks, like agriculture or manufacturing. Mencius criticizes them as "sprout pullers," ignorant farmers trying to force a rationalist morality on people, ignoring innate tendencies and thereby doing nothing but harm in the end. He similarly argues against those Confucians of his own time who advocated a radical reshaping of the self. Although Mencius sees Laozi's endorsement of the "uncarved block" as an excuse for lazy backwardness, he is equally critical of the carving strategy for attaining *wu-wei*, as we see in a famous exchange with a person named Gaozi. There is scholarly debate about who this Gaozi was. Evidence from some recently discovered archaeological texts suggests that he, like Mencius, was a follower of Confucius, the difference being that Gaozi advocated a more authoritarian, effort-based reshaping of the self—the carving-and-polishing strategy. (In other words, as a historian of religion, I'd be inclined to say that Gaozi, like the later Xunzi, was a *real* Confucian, as opposed to whatever we want to call Mencius.)

In a series of exchanges, Gaozi proposes metaphors for self-cultivation that suggest a need for effort and a reshaping imposed from the outside. He begins with the promising carving metaphor: "Human nature is like a willow tree, and moral behavior is like cups and bowls. To make people moral is like making cups and bowls out of the willow tree." Confucius would approve! Becoming good is like carving and polishing jade or bone. Mencius, however, is not at all satisfied: "Can you follow along with the nature of the willow when you make it into cups and bowls? Or is it rather the case that you have to mutilate the willow before you can carve it up into such shapes? If you have to mutilate the willow to make it into cups and bowls, must you also mutilate people to make them moral? All that your strategy will achieve is to make people bring disaster and confusion upon themselves!" Fair enough. Gaozi abandons the carving metaphor and switches to a water metaphor, trying to drive home the point that humans have no particular inclination toward or against morality: "Human nature is like an irrigation pond. Open the gate to the east and the water will flow east; open the gate to

the west and the water will flow west. Water doesn't have a preference for flowing east or west; similarly, human nature has no tendency toward good or bad." Here the Confucian educator uses his
cold cognition to determine the proper direction for human behavior and then trains hot cognition appropriately. Again, this wouldn't
be out of place in the *Analects*, but Mencius still objects: "Water certainly has no preference for east or west, but is it also indifferent to
up versus down? The goodness of human nature is like the downhill flowing of water—there is no person who does not tend toward
the good, in the same way that there is no water that does not flow
downhill."

In a kind of metaphorical jujitsu, Mencius is twisting Gaozi's
images around to demonstrate the futility of Gaozi's carving-and-
polishing strategy. For Mencius, the only way to get into *wu-wei* is
to tap into innate tendencies and then cultivate them until they are
strong enough to take over. If you can reach this point, socially desirable behavior will flow from your hot cognition like water flow-
ing downhill. The result is the same kind of flexibility and cultured
grace that we saw in the *Analects* of Confucius, but one originating
from, and organically grounded in, our biological selves.

One implication of this organic grounding is that the cultural
forms inherited from our ancestors become less crucial in attaining
wu-wei. Although, as a Confucian, Mencius was dedicated to the ancient Zhou rites and classics, we see in the *Mencius* hints of a kind
of cultural liberalism that would bother Confucius. In one passage,
Mencius declares that it would be better to not have the sacred Book
of History at all than to believe everything in it; if there is a tension
between what the classics tell you and what your sprouts say, go with
your sprouts. Traditional Confucian culture is demoted from an essential tool in the shaping of a properly formed human being to a
useful guide—what the scholar of Chinese thought P. J. Ivanhoe has
called a kind of "trellis" for one's moral sprouts. A trellis may help
the sprouts to grow faster or produce more fruit, but it's not indis-
pensable.

There are, in fact, several instances in the text where we find Mencius playing fast and loose with traditional norms. For instance, in his efforts to reform King Xuan of Qi, Mencius apparently spent a lot of time talking to the king's ministers—which, you'll recall, is how he first heard about the sparing of the ox incident. In one passage, we find him similarly encouraged by a report claiming that the king has expressed an interest in music. As with the ox incident, Mencius confronts the king with what he has heard: "Is it true that you are fond of music?" The king blushes and admits, "It is not the music of the Ancient Kings that I enjoy, but rather contemporary popular music." In other words, the king prefers Bon Jovi to Beethoven. In fact, given his documented predilections—he likes women, wine, and a good party—my guess has always been that the king is a fan of the famously licentious music of Zheng that was condemned by Confucius.

We know what Confucius's response would be: Give up this horrid, decadent music, and cultivate a taste for the classics. This is not, however, Mencius's reaction. "If you have a fondness for music, then perhaps there is hope for the state of Qi," Mencius observes. "Whether it is contemporary music or the music of the ancients makes no difference." Huh? The king himself is surprised; he obviously expected a lecture. Relieved not to be scolded, he eagerly asks, "Can I hear more about this?" Mencius then proceeds to explain how, no matter what type of music the king enjoys, he would enjoy it even more if he had a clear conscience about how he treated his people. He leads the king through another visualization exercise. Imagine you are in your palace enjoying your music, but outside the people are grumbling about the expense and wondering how you can take time to listen to music when they are all suffering. Doesn't that decrease your enjoyment? Now imagine that you make some effort to be a better king: maybe you cut your brutally high taxes, or stop pulling farmers out of the fields at harvest time to work on your lap pool. Now when the people hear the sounds of music drifting over your palace walls, they will smile at one another and think, "Ah, our wonderful king is in good health and fine spirits. How nice! We hope he's having a good time." Wouldn't that be better? Wouldn't you then enjoy your wine, women, and song even more?

This passage comes after the exchange about the ox, and it's always struck me that, by this point, Mencius appears to have given up on the whole sprout thing with King Xuan. He seems resigned to going with whatever might work to make the king's rule a bit less horrid for his people. It's significant, though, that even with such a seriously debauched case as the king, Mencius resists trying to impose behavioral restrictions upon him or shaming him into thinking that his current habits are bad and need to be reformed. It's clear that Mencius thinks that forcing things is anathema, whether we're talking about Mohist-style rationalism or Confucius-Xunzi–style cultural training. We can see that Mencius really seems eager to split the difference, as it were, between Laozi and Confucius by presenting *wu-wei* as the natural outgrowth of cultivating our nature. It's a promising model for how you could direct spontaneity in such a way that you don't destroy it in the process.

WHY IS BEING "NATURAL" SO MUCH WORK?

Confucius faced a problem when it came to motivation and his carving-and-polishing strategy. Training in ritual and the classics works only if the student has some sort of incipient appreciation of the Confucian Way. To master the Way, you need to, at some level, already *love* it. According to Confucius's own assessment, however, very few of his contemporaries felt this love. Most of them would rather pursue wealth, physical comfort, or sex than memorize the *Book of Odes*—an attitude I think many of us today can sympathize with. How do you change this? How do you get someone to love something that they don't already love?

Mencius "solves" this problem by saying we *do* already love the Way. Human nature is good, he famously declares: we have within us incipient versions of all of the Confucian virtues, in the form of the sprouts, which are eager to grow if we can just help them. Mencius works hard to show even hard cases like King Xuan of Qi that these sprouts of virtue can be discerned beneath his callous and lazy surface. If he can focus on that fragile little sprout of virtue, and nurture and strengthen it, it will naturally grow into an *wu-wei* compassion powerful enough to make him like the true kings of ancient times.

One might be forgiven for wondering, though, why we need Mencius to remind us what our true nature is. We seem perfectly capable of getting in tune with our desires for pleasure and comfort without the need for Mencian intervention. People like King Xuan seem to naturally, spontaneously prefer eating, drinking, hunting, and womanizing to behaving compassionately. He is, in fact, an *wu-wei* party animal. Mencius is essentially trying to tell him that the things he *seems* to like are not what he really desires deep down; in pursuing a life of sensory indulgence, he is mistakenly giving his "petty parts" (his mouth, his stomach, and other regions we won't mention) preference over his noble parts (his mind and his moral sprouts). If this is true, though, why do so few people seem to spontaneously follow their noble parts?

This tension is brought to a head by one of Mencius's disciples, who approaches the master because he is having doubts about his own moral potential. He has heard from Mencius that everyone has within them the propensity to become a perfectly *wu-wei* moral exemplar. However, he finds that, when left to his own devices, all he really feels like doing is "sitting around and eating millet all day long"—the early Chinese equivalent of sprawling on the couch, eating potato chips, and watching reality TV. Mencius is unconcerned:

What is the difficulty? All you have to do is *try* at it! Is the trouble with people that they do not have the strength? No, it is that they do not try, that is all. One who walks slowly behind his elders is called a proper young man, while the opposite is true of one who walks quickly and overtakes his elders. Now, is walking slowly really something people aren't capable of? No, it is just that they don't try. . . . The Confucian Way is like a wide road. How is it hard to find? The trouble with people is simply that they don't seek it out.

As in his exchange with King Xuan, Mencius's reply to those who profess to lack moral motivation is to say, no you don't! Morality just seems *so hard*, they reply. No matter, just do it! People have the Way within them, and if they don't find themselves being drawn to developing it, it's only because they are not making an effort to seek it out.

This is an unsatisfying explanation. When I am stuck walking behind some creaky, slow-moving old geezer, it is certainly within my physical powers to slow my pace and respectfully allow him to take the lead, but it's not my inclination. I have places to be, things to do, and I generally slip past the first chance I get. When in China, where such behavior is—or at least used to be—somewhat frowned upon, I *try* to be more patient and polite, but it feels like effort, not *wu-wei*. Mencius "solves" the motivational tension we saw in the *Analects* by claiming that we already love the Confucian Way, we just need help realizing it. However, one is led to wonder about the naturalness—or even existence—of a moral "love" that is so hard to see and that takes so much outside guidance and work to feel. The fact that the moral sprouts are rather elusive suggests that they are not as strong or innate as Mencius claims. Therefore, despite apparently having been vanquished by Mencius's sprout-cultivation strategy, the basic motivational tension we saw in the *Analects* is very much alive and well. We're being asked to love something we don't seem naturally inclined to love.

Thinking about it in terms of Mencius's agricultural metaphor, we might note that there is nothing at all natural about growing crops. As anyone who has ever tried it can attest, it's a huge amount of work. Crops are not natural, *weeds* are. One might then reasonably conclude that the real secret to *wu-wei* might be found among the weeds of humanity who live and flourish outside the boundaries of the carefully tended Mencian fields. The key is not to focus and cultivate but to let the world take you where it will.

6

Forget About It

GOING WITH THE FLOW

THE EARLY DAOIST THINKER ZHUANGZI—THE PURPORTED AUTHOR of the text bearing his name—kept some odd company. The *Zhuangzi* is full of talking animals and insects, mysterious leviathans that transform into huge birds, witches, hunchbacks, ghosts, talking skulls, and ancient sage kings come back to life. Perhaps one of the stranger aspects of the *Zhuangzi*, though, is the friendship that it documents between Zhuangzi and a man named Huizi. Huizi was also a famous thinker of the time, although his own writings did not survive. On the basis of his appearance in the *Zhuangzi*, and some fragmentary teachings attributed to him in other early texts, Huizi seems to have been a Mohist whose specialty was logic. Huizi believed that, by carefully defining one's terms and perfecting the technique of logical argumentation, one could demonstrate the rational superiority of Mohist utilitarianism. Of course, as a Mohist, he thought that was all you needed to do: convince people rationally that they *should* be utilitarians, practicing impartial caring toward all, and they'd start behaving that way. For a thinker like Zhuangzi who believed that language and logic were dangerous traps, Huizi seems like an unlikely best friend.

To be sure, Huizi typically serves as Zhuangzi's straight man, the plodding egghead who just can't seem to understand the deficiency of reason alone and who is blind to the power of *wu-wei*. He always loses their arguments and is sometimes made to look a bit silly, but there is clearly genuine affection between these two men. One typical exchange begins with Huizi telling Zhuangzi that a king once gave him a gift of a handful of large gourd seeds: "When I planted them they grew into enormous gourds, big enough to hold twenty gallons! I tried to use them as water containers, but they were too heavy to lift; I tried cutting them to make spoons, but they were too shallow to hold any liquid. It's not that I wasn't impressed by their size, but I decided they weren't really useful for anything, so I smashed them."

In China at the time, gourds were used for these two purposes, containers or spoons. Hence Huizi's disappointment. Hearing this story, though, Zhuangzi is incredulous. "You are certainly a clod when it comes to thinking big!" he declares. He tells Huizi some stories about people who took apparently useless or trivial items and used them for unexpected purposes, winning great rewards in the process. "Now you've got these gourds," he concludes. "Why didn't it occur to you that you could turn them into a big raft to float around on the rivers and lakes, instead of lamenting how they're too big to use as spoons! It's as though you've got underbrush growing in your mind!"

The problem with Huizi is that he can't see beyond the possibilities defined by his culture. Gourds are used for X or Y; these gourds do not work for either purpose; ergo, they are useless. Psychologists refer to this as "categorical inflexibility," a tendency for socially learned representations of objects to constrain our ability to think about them in novel or creative ways. This mental inflexibility is a symptom of a problem that Zhuangzi saw as the primary barrier to *wu-wei*: the human tendency to be dominated by the "mind"—a term that, for Zhuangzi, refers to what we'd think of as the cold-cognition system. "Letting the mind be your master," as he puts it, causes people to get trapped in rigid social categories, artificial values, and narrowly instrumental reasoning, preventing them from seeing the world clearly and getting into its flow.

Mohists like Huizi are certainly part of the problem. With their confidence that proper behavior can be guided solely by logic and calculation, they completely ignore the spontaneous wisdom of the body. The Confucians aren't any better, though. With their preestablished ideas about what spontaneity should look like, and their rigid techniques for cultivating it, they end up being as alienated from the world and their own authentic feelings as the Mohists. Indeed, Zhuangzi often portrays the Mohists and Confucians as a kind of philosophical Tweedledum and Tweedledee, each side declaring that its Way is absolutely right, but neither side being able to see its failings.

Zhuangzi felt that this left the people of his age in a terrible mess. One mitigating factor with the Mohists and Confucians is that, despite their ineffectiveness, they at least want to help others and improve the world. The problem is that their "petty understanding"—Zhuangzi's term for any narrow, rigid view of the world—sets the tone for society as a whole, giving rise to a society composed of arrogant know-it-alls, publicly prosperous and confident but secretly miserable, feverishly pursuing a false vision of happiness that always lies just beyond their reach. A haunting passage from early in the text contrasts the "great understanding" that Zhuangzi would like us to embrace with the spiritual bankruptcy and suffering that he saw all around him:

> When people are asleep, their spirits wander off; when they are awake, their bodies are like an open door, so that everything they touch becomes an entanglement. Day after day they use their minds to stir up trouble; they become boastful, sneaky, secretive. They are consumed with anxiety over trivial matters but remain arrogantly oblivious to the things truly worth fearing. Their words fly from their mouths like crossbow bolts, so sure are they that they

know right from wrong. They cling to their positions as though they had sworn an oath, so sure are they of victory. Their gradual decline is like autumn fading into winter—this is how they dwindle day by day. They drown in what they do—you cannot make them turn back. They begin to suffocate, as though sealed up in a box—this is how they decline into senility. And as their minds approach death, nothing can cause them to turn back toward the light.

This is an eloquent but grim vision. As an analysis of the problems at the heart of our modern, striving society, it's hard to beat—which makes it all the more amazing that is was written two thousand years ago in classical Chinese. The fact that this passage is actually targeted at the desperate inhabitants of Warring States China suggests that the challenges of finding happiness in civilized life have not changed much over the millennia. One can't help but think of ruthless corporate climbers, sacrificing their youth, their health, their family to make it to the top, only to find once they reach the corner office that they're too exhausted and dispirited to enjoy it. Thoughts also turn to wealthy suburbanites, straight out of *Desperate Housewives*, endlessly accumulating more and more possessions, bigger houses, and fancier cars; running the treadmill; taking Pilates; gossiping at the club; but ultimately plagued by a vague feeling of meaninglessness. The way off the hamster wheel, according to Zhuangzi, is to *stop* trying harder, learning more, and laboriously cultivating the self. We need to learn to let go. Once we can do this, we will be truly open to the world and to other people, and *wu-wei* will come to us naturally.

FORGET, LET GO

The tone of Zhuangzi's critique of Warring States society, and his suspicion of both cold cognition and explicit morality, sounds more than a little like what we've seen in the *Laozi*. Socially learned values lead us astray; knowledge is dangerous; forced morality is not moral; go with your belly, not your mind or your eye. This is why, when later librarians were given the task of cataloguing the imperial library, the *Laozi* and the *Zhuangzi* were grouped together as "Daoist" texts, a characterization that persists to this day. It's important to realize, though, that—unlike the Confucians—the "Daoists" were not self-identified members of a formal school, and behind their many similarities remain important differences. For instance, despite his concerns about artificial, consumerist, rat-race society, Zhuangzi didn't think that running away was the answer. He thought that retreating to a self-consciously primitive lifestyle was just as misguided as being a full-blooded Confucian or Mohist, because it involved setting up a concrete "right" way of life in contrast to a rejected "wrong." Zhuangzi would have made fun of the 1960s hippies as well as their parents.

The key, in his view, was to not condemn others or to pride yourself on being right but rather to get beyond right and wrong altogether: "If you're committed to something being 'right,' you're equally committed to something else being 'wrong'; condemning something as 'wrong' means valuing something else as 'right.' This is why the sage does not go down this road but simply illuminates things by means of Heaven. He still follows a 'this,' but in such a way that his 'this' is also a 'that,' his 'that' is also a 'this.'" We find Zhuangzi's exemplars living more or less normal lives, often employed in eminently civilized activities. The stories of Butcher Ding and Woodcarver Qing come from the *Zhuangzi*, and they are firmly embedded in mainstream Confucian culture, the butcher cutting up oxen for ritual sacrifice, Qing carving bell stands for a classical musical performance.

Zhuangzi's exemplars are, however, distinctive in a couple ways. To begin with, they are a diverse bunch. While Confucius and Mencius hobnobbed with lords and kings, Zhuangzi was hanging out in workshops and the kitchens—from which, as Mencius sniffed, the gentleman "keeps his distance"—and he was impressed by what he saw. This world revealed to him artisans and butchers, ferrymen and draftsmen, whose effortless ease and responsiveness to the world could serve as a model for his disaffected fellow intellectuals. His debates with Huizi were aimed at getting his friend out of his head and into this world, where he could follow his body and not his mind. In keeping with Zhuangzi's refusal to be confined by standard social values, his sages include hunchbacks, lepers, witches, and criminals with amputated feet—the "weeds" of humanity ignored by Mencius with his carefully tended moral garden, but more representative, in Zhuangzi's view, of true naturalness.

Zhuangzi's sages are also distinctive by virtue of the fact that they are not attached to strict values. They live their lives and have their goals but maintain an openness that allows them to change direction when circumstances demand, or to let go of something that has turned from a gift to a burden and move on to something else. Translated into modern terms, they've downregulated their cold cognition so that their hot cognition can run the show more or less directly, with minimal interference from the conscious mind. The trick, of course, is how to pull this off.

What does Huizi need to do to be less cramped by social norms, to respond to the world as it is, not as he thinks it should be? He needs to forget and let go. How does he do *that*? He needs to escape the domination of the conscious mind. This is laid out most clearly in a famous passage where—in a typically facetious move—Zhuangzi recounts an exchange between Confucius and his favorite disciple, Yan Hui, but with Confucius actually espousing very Zhuangzian (and quite un-Confucian) ideas. Yan Hui has heard about a ruler in a neighboring state who has been oppressing his people, and he decides he's going to visit this guy and set him straight. Confucius is doubtful that he will succeed or even return with his head still attached to his shoulders. The problem seems to be that Yan Hui is motivated by the wrong things: abstract teachings that he has heard, his arrogance in thinking that he is wiser and better than the ruler, and—as Confucius rather sharply suggests—a not-so-hidden desire to achieve fame as the hero who reformed the evil king. Yan Hui suggests several different strategies, all of which are shot down by Confucius. "You are still being controlled by your mind!" Confucius complains repeatedly. Finally Yan Hui gives up. "I have no further suggestions," he sighs. "May I ask what you think the proper way of going about this might be?"

"You must fast!" Confucius responds. "Let me tell you, you'll achieve nothing as long as you're listening to your mind. Those who are guided only by their minds are not considered fit by Bright Heaven."

Yan Hui replies, "My family is poor, so I haven't drunk wine or eaten meat for several months. Is that what you mean by fasting?"

"No! That is the kind of fasting you do before a sacrifice. What I am talking about is the fasting of the mind."

"What is that?"

Confucius says, "Unify your intentions. It's better to listen with your mind than to listen with your ears, but better still is to listen with your *qi*. The ears only record sounds, the mind can only analyze and categorize, but the *qi* is empty and receptive. If you make yourself empty, nothing less than the Way itself will appear to you. This emptiness is what I mean by the fasting of the mind."

Even in the West many are by now familiar with the concept of *qi* (pronounced *chee*, also known in the alternate Romanization *ch'i*), seen by the early Chinese as a kind of vital force that animates all living things. By the time that Zhuangzi was writing in the fourth century B.C.E., the meaning of the term *qi* had come to encompass as well a substance that provides people with a direct connection to their true or Heavenly natures. It's a distinctly religious view of *qi*, one that links it to a kind of sacred power within the self: the "spirit," as in the "spiritual desires" that guide Butcher Ding through the ox. Because of its connection to Heaven, it also gives unique and direct access to the Heavenly makeup of things, as we saw in the Woodcarver Qing story, where the artisan's ability to escape the domination of his mind allows "the Heavenly within to match up with the Heavenly in the world."

One way to understand the distinction in this passage between the three levels of "listening" or perceiving the world is to see them as engaging different areas of the brain. Listening with the ears is a bit like Butcher Ding merely looking with his eyes at the huge ox in front of him: he's just taking in sensory information but has no idea what to do with it. Listening with the mind involves regions like the lateral PFC, which consciously analyzes this information and relates it to prior knowledge. Listening with the *qi* seems to refer to shutting down the cognitive control regions of the brain—what we think of as the conscious mind—and letting the adaptive unconscious take over. In the context of the early Chinese worldview, this unconscious is going to lead us in the right direction because it possesses a sacred quality. Like the "spiritual desires" in the Butcher Ding story, this *qi* is a force connected directly to Heaven. Indeed, for Zhuangzi the spirit and the *qi* seem to be more or less synonymous: both provide one with a pipeline to Heavenly guidance.

However we're to understand this advice about emptiness, it has an immediate and powerful effect on Yan Hui, inspiring a kind of instant enlightenment. "Before I was able to grasp this teaching, I was full of thoughts of myself," he declares. "But now that I get it, it's as if my self has never actually existed. Can this be called emptiness?" "You've got it!" the Master answers. "Let me tell you: now you can go and play in his gilded cage without being moved by fame. If he listens to you, then sing; if not, remain silent. Forget everything you've been taught and abandon all personal schemes. Reside in oneness and lodge yourself in what cannot be stopped. Then you'll be close to success."

The loss of self that Yan Hui reports involves abandoning all self-serving strategic thinking and preconceptions. By emptying himself of himself, as it were, he creates a receptive space, an openness to hearing what the ruler actually has to say and what the situation actually demands. He shuts down his conscious, calculating mind and lets his vital energy or *qi* take over. He thereby becomes absorbed into something larger than himself: the movement of the Way, that which "cannot be stopped," a sacred force that will carry him along to the proper outcome. Plodding, rigid, conscious planning is replaced with fast, flexible, and unconscious responsiveness to the world.

More than anyone else in Warring States thought, Zhuangzi perceived the limits of conscious thought and celebrated instead the unique powers of the embodied mind. We discussed in chapter 1 how important these hot processes are in terms of directing the bulk of what we do and how effective they are in terms of moving us through the world. It's also becoming increasingly clear that the kind of cognitive flexibility that Zhuangzi saw as so fatally lacking in his contemporaries is something that is best achieved when we can weaken the hold of the conscious mind.

This is easier for children than adults. Take "divergent" creativity, which refers to the ability to imagine multiple solutions to a problem, or novel uses for an object. A common way to measure it experimentally is the Unusual Uses Test (UUT), where subjects are given a common object and asked to come up with as many different possible uses as they can imagine within a given time frame. Young children are more flexible and creative in such tests, not only because they've had less time to be indoctrinated in what these objects "are for," but also because their cognitive control regions are less developed. Kids immediately grasp that Huizi's huge gourds would make an awesome raft. Interestingly, successful performance by adults in tasks that require creative recategorization is accompanied by measurable downregulation of the cognitive control regions, and adults with PFC damage tend to do better on such tasks than healthy controls.

Alcohol, a very effective means of temporarily paralyzing our cognitive control abilities, has also been shown to enhance various types of creativity. One recent study asked subjects to perform a Remote Associates Test (RAT). In the RAT, subjects are given three apparently unrelated words (say, *peach, arm,* and *tar*), and asked to come up with a fourth word that will connect them (in this case, *pit*). Although the RAT is often used to probe "convergent" thinking—which seems distinct from "divergent" thinking and requires more cognitive control—it's become clear that when initial guesses are incorrect, or subjects just can't "see" the solution immediately, divergent thinking becomes crucial. In the study in question, the researchers found that subjects brought to a moderate level of inebriation—a blood alcohol level of .075, just shy of where you'd lose your driver's license—did better on the RAT than sober controls. Moreover, the inebriated subjects were more likely to attribute their success to sudden "insight" rather than more plodding analytic strategies. Getting a bit drunk seems to weaken cognitive control and enhance insight-based creativity.

Similar effects have been found in "incubation" experiments, where subjects are given a primary task but then are allowed some time off, as it were, distracted by another task. As long as the primary task goal is kept salient—that is, as long as it remains somewhere in the back of the mind—brief distractions appear to enhance both problem-solving ability and skilled performance on physical tasks. Again, this seems to be because hot systems are good at making cognitive leaps. If the conscious mind can be temporarily diverted, the unconscious mind is free to get on with its work.

Recall the "drunk on Heaven" passage, where the inebriated cart rider emerges unscathed from a tumble because his "spirit is intact"—his cold cognition is taking a little rest and his hot processes are running unimpeded. Not being troubled by the conscious fears that plague us sober people, he can roll with the fall and avoid injury. We see a similar theme in another imaginary dialogue between Yan Hui and Confucius. Yan Hui reports that he's just had a hair-raising experience, crossing a dangerous and turbulent river with a ferryman who seemed completely fearless and who handled the small boat with nearly supernatural skill. When Yan Hui asked the ferryman his secret, the ferryman replied that anyone able to swim could do what he did. (In many traditional societies learning to swim is a rare achievement, and crossings of water are therefore always fraught with anxiety.) Confucius, again serving as Zhuangzi's mouthpiece, replies that someone who swims well has "forgotten the water"—that is, they no longer fear it, and so it no longer takes up space in their consciousness. As a result, they can easily learn to master a small boat because their freedom from fear and distraction allows them to just relax into their unconscious skill: "They look upon the vast deep as if it were safe, dry land and view the capsizing of a boat with the same equanimity that you would view your cart tipping over. Imagine viewing the upheavals and setbacks of everyday life with the same lack of concern—nothing could get in and bother you, and there would be nowhere you could go and not be at ease." Confucius then uses the example of an archery contest to illustrate the harmful effect of consciously focusing on extraneous concerns: "If you're betting for broken bits of potsherd you can shoot with perfect skill—there is nothing at stake. If you start betting for belt buckles, you become worried about your aim. By the time you start betting for gold, you're completely petrified. Your actual skill is the same in all three cases, but because of the relative value you place on these objects, you end up paying more attention to extraneous things. It is always the case that those who focus on the outside become clumsy on the inside."

We couldn't ask for a better summation of the problem facing the athletes and performers we met in the Introduction. Such people have fallen out of their usual, *wu-wei* immersion in the internal goods of the game; they have become alienated from the goals, the values, and the flow of play. It is not that their actual physical skill has changed, it's that they've allowed concern with externalities to make them "clumsy on the inside."

Zhuangzi's insights in this regard find confirmation in what is now a fairly extensive psychological literature on the phenomenon of "choking." The consensus is that, in most cases, external pressures—explicit demands for good performance, concerns about reputation or awards—cause people to consciously focus on activities that should be handled by the unconscious. The result is disruption, or "paradoxical performance": the more you try, the worse you do. In one classic experiment (the date of which, 1984, can be roughly surmised by the technology employed), experienced players of Pac-Man or Ms. Pac-Man displayed a significant decrease in performance when monetary rewards were introduced. A more recent study of expert baseball players showed that, when subjected to high pressure, their performance suffered. Significantly, though, their conscious awareness of what they were doing—for instance, how their bat was oriented at any given time—was enhanced. Like Yan Hui when he first plans to go advise the evil king, choking baseball players are letting themselves be controlled by the conscious mind, with disastrous results.

According to Zhuangzi, one of the worst external distractions is the kind of explicit morality taught by the Confucians, along with their rigid strategies for achieving *wu-wei*. This is highlighted in another Confucius–Yan Hui dialogue—one that recalls the "fasting of the mind" passage we discussed earlier but that appears a couple chapters later in the *Zhuangzi*. Presumably Yan Hui has once again told Confucius about his arrogant, misguided plan to go reform a corrupt king, and Confucius has shot him down. This time, however, Confucius doesn't give him any specific advice but just sends him away to do something—what exactly, we are not told. Twice Yan Hui pops back in to declare, "I'm improving!" and to update Confucius on his progress. The first time he reports that he has forgotten benevolence and righteousness, the two most important Confucian virtues. The second time he reports that he's forgotten Confucian ritual and music. "That's not bad," Confucius replies both times, "but you are still not there." Things go a bit better on his third visit:

They met again on another day, and Yan Hui said, "I'm improving!"

"What do you mean by that?"

"I can sit and forget everything!"

Confucius looked surprised and said, "What do you mean, sit and forget everything?"

Yan Hui replied, "I let my four limbs and body fall away, drive out perception and thinking, separate from my physical body, get rid of knowledge, and harmonize myself with the Way. This is what I mean by sitting and forgetting everything."

Confucius replied, "Being harmonized, you must be free of preferences; having been transformed, you must be free of rigidity. So you really are a worthy man after all! I humbly request to become *your* disciple."

As with the Woodcarver Qing story, we have the process of "forgetting" taken to an extreme: not only the moral teachings and practices of the Confucians but also the body and perception itself. The idea here is that, if you are to successfully enter *wu-wei*, your focus should be on the world, not yourself. You have to forget everything—your ego, even your own body—so that you can be absorbed into the larger movement of Heaven's Way.

Sounds great. The question, of course, is how *do* you lose yourself? How do you transfer control from the conscious mind to the body? More specifically, what the heck is Yan Hui *doing* when he goes away to "sit and forget"?

It's possible that Zhuangzi dabbled in the sort of meditation and breathing techniques that seemed part of the Laozian strategy, and there are tantalizing hints that psychoactive substances were also involved. For instance, one of the early chapters of the book opens with a description of a certain Ziqi of South Wall, who is sitting in an odd, reclined posture, occasionally raising his head to the ceiling and taking deep breaths, looking "vacant and dazed, as if he had lost his companion." Who is his "companion"? Your guess is as good as anyone's, although most commentators think it refers to his body. Anyway, this guy has been doing intense meditation practice or serious drugs—or both. It gets weirder: he has an assistant, who apparently has been standing in front of him in attendance, watching or helping out in whatever's been going on. (Helping him time his breathing? Ready to intervene if Ziqi starts to have a bad trip? Again, we don't know.) Apparently the assistant is impressed: "Wow, what just happened? Is it really possible to make your body like dry wood, to make your mind like dead ashes? The person sitting there now is not the same person who was sitting there before."

Ziqi confirms that he's experienced a profound transformation. "My friend, that's a very good question. Just now I lost myself. Can you comprehend that?" When asked to elaborate, he goes on for a long time trying to explain what he just experienced. There's all this crazy stuff about the way the wind howls when it blows over the landscape, the pipings of Heaven and Earth, and questions about where the wind comes from. The experience has apparently done Ziqi some good, though, because we then immediately segue into the passage about the "great understanding" versus the "petty understanding"—we're led to conclude that Ziqi's experience has turned him into a sage.

The figure of Zhuangzi is associated historically with the southern state of Chu, which corresponds roughly to present-day Hubei and Hunan provinces, located in the center of present-day China. In the Warring States period, however, Chu was a border state, portrayed as an exotic, semibarbarian realm full of strange foods and wild animals. It is also traditionally associated with shamanistic practices, including astral projection trips and trances. Moreover, the stylistic transition as one moves from fascinating, but relatively staid, earlier texts to the *Zhuangzi* is nothing less than shocking. All of a sudden we have talking animals and people flying around or getting reincarnated as rat's livers. One of the difficulties of the text is that Zhuangzi is forced to invent whole new sets of adjectives and adverbs to get across the experiences he's trying to convey, which simply couldn't be captured in the classical Chinese of his day. Moving from the *Mencius* to the *Zhuangzi* is not unlike moving from the Beatles' early music—for instance, "Can't Buy Me Love"

(1964)—to "Sgt. Pepper's Lonely Hearts Club Band" (1967): you get the distinct impression that somebody's just discovered drugs.

The fact remains, though, that Zhuangzi never actually tells us what his sages are up to, and that's probably no accident. Prescribing a course of hallucinogens, setting up a specific meditation regimen, or giving us a set of breathing exercises would land him in the trap of setting up a "right" in opposition to a "wrong"—the worst thing you can do. Because he wants to avoid all rigid claims about good or bad, Zhuangzi is limited to simply holding up images for us to consider: those in the thrall of "petty understanding," drifting miserably into senility, darkness, and death; Butcher Ding or Woodcarver Qing moving smoothly and easily through their lives; Huizi smashing up some wondrously large gourds and thereby missing an opportunity to drift happily around the rivers and lakes. Obviously, some of these images portray "good" ways of life, and others "bad," but Zhuangzi will never come right out and say that. He tells us stories and lets them have their effect.

In my view, this is probably his strategy for getting people into *wu-wei*. He is trying to use humor, paradox, or simple weirdness to shock us out of our normal ways of thinking. The text of the *Zhuangzi* is trying to *do* something to you, and its effectiveness in that regard is difficult to convey—you need to experience it. Isolated quotations can only give a sense. Consider this passage, for instance: "There is a point at which something begins. There is a point at which we've not yet begun to have a beginning. There is a point when we've not yet begun to begin to not yet have a beginning. There is existence. There is nonexistence. There is also a point where we have not yet begun to have nonexistence, and when we have not yet begun to have a beginning to nonexistence." It goes on like this for a while. Some traditional Chinese commentators and modern scholars of the text have racked their brains trying to figure out the logic involved in these statements, treating this passage as if it were a straightforward philosophical argument. It's almost certainly not: the skepticism about our possibility of knowledge expressed here is meant as spiritual *therapy*, not religious doctrine. Reading it, our certainty that logic and rationality can get us to where we want to go is shaken, and that's the effect Zhuangzi wants to achieve.

This strategy of using language to undermine itself was enthusiastically adopted by the Chan/Zen Buddhists, the branch of East Asian Buddhism most directly and pervasively influenced by Zhuangzi. They formulized this technique into a practice referred to as *gong-an* (literally "public case") in Chinese, but better known in the West by the Japanese pronunciation *koan*. *Koans* consist of riddles, nonsense statements, or interpersonal encounters that either have no logical meaning or try to unsettle our ideas of what Buddhism is about. They are to be meditated upon in order to break the hold of rationality on the self—to "fast away the mind." A common structure to the encounters is that a student, trying to be a good Buddhist, asks some straightforward question about doctrine. The Zen master then answers with a non sequitur, trying to shake up the student's conceptual framework. The student then gets confused, hesitates, maybe begins to ask another question, and then the master strikes (verbally or physically) in order to provoke something like a spiritual nervous breakdown. The goal is to mount a multi-pronged assault on ordinary reason—verbal, physical, social—in order to free the embodied mind from the limitations of cold cognition and shock the student into a state of *wu-wei*.

Many of these encounters involve radically unconventional behavior (wearing sandals on one's head, going out naked) or physical violence of some sort. There is a lot of slapping, hitting, and whacking of people with sticks. Cats get cut in two. The most compelling techniques, though, are a little more subtle and try to direct students' attention away from abstract doctrines or future plans and toward the reality that is right in front of them. One famous *koan* story, for instance, begins like this: "A monk said to the Zen master: 'I have just entered the monastery. I beg you, teacher, to provide me with instruction.'" This monk sounds a lot like the pre-fast Yan Hui: an eager Boy Scout, with clear expectations about what he is to learn, and probably a subtle desire to demonstrate his ambition. A *keener*, as we would say in Canada.

> The master asked, "Have you eaten your rice porridge?"
> The monk replied, "Yes, sir, I have."
> The master said, "Then go wash your bowl."
> At that moment the monk was enlightened.

The key to enlightened *wu-wei* is not learning more about doctrine but seeing and responding appropriately to what is in front of you. The Chinese Zen master who compiled the collection in which this *koan* appears comments, "It is only because it is so clear that it is so hard to see. People go looking for fire using a lighted lamp; if they only realized that the lamp itself was fire, they'd be able to cook their rice much sooner."

WANDERING FREE AND EASY

Like the enlightened monks in the *koan* literature, the exemplars that appear in the *Zhuangzi* are designed to show us how happy and effective we would be if we could transfer control from the conscious to the unconscious mind. Although Zhuangzi often uses the activities of artisans or workmen to illustrate this point, the examples of Butcher Ding and Woodcarver Qing are meant as metaphors—the real goal is to teach us the "secret to living life." Butcher Ding's blade has no thickness, so it can play in the spaces in between the tendons and bones. In the same way, a person who genuinely has no self can move smoothly through the social world.

Zhuangzi shares with our other thinkers a conviction that *wu-wei* leads to *de*, although the power of Zhuangzian *de* lies not so much in attracting others as in relaxing them. At one point Zhuangzi notes that if you are out boating on a lake and someone in another rowboat collides with you, you get angry, yell at them, and curse. If the same thing happens with an empty rowboat—say, one blown into your path by the wind—you simply shrug it off and go on your way. The goal of the Zhuangzian sage is to empty his boat, so he can collide with others without arousing any animosity.

There are also suggestions that, should the Zhuangzian sage nonetheless attract attention to himself, his powerful *de* will be clearly visible to others, allowing him to proceed on his way unmolested. One story tells of a man who is training fighting roosters for a king. After ten days of training, the king asks if they are ready.

> "Not yet. They are still too arrogant and dependent on their physical *qi*."
>
> After another ten days the king asks again.
>
> "Not yet. They still react to noises and movements."
>
> After another ten days the king asks again.
>
> "Not yet. They still look around aggressively and are overflowing with *qi*."
>
> Finally, after another ten days, the king asks again.
>
> "They are close enough. Even if another rooster crows they show no change of expression. Looking at them, you'd think they were made out of wood, so perfect and complete is their *de*. Other roosters will not even dare to approach them, and just turn and run."

Again, Zhuangzi's aim here is to tell us, not how to run a successful cockfighting business, but how to live a successful life. Complete relaxation and freedom from external concerns perfect your *de* and make you formidable, conveying a confidence and ease that makes others think twice before messing with you.

Yan Hui at the end of the "fasting of the mind" story is a great example of how selflessness, interpersonal responsiveness, and social success go together. No longer deluded by rigid teaching or preconceived goals, and free of ulterior motives and self-importance, he is able to *really* listen to the king and to "sing" only when the king is ready. In this way, he'll not only manage to keep his head attached to his shoulders but also potentially win over the king. We see a similar theme in a story about a monkey trainer: "There was a monkey trainer who was distributing nuts to his monkeys, saying, 'You will get three in the morning and four in the evening.' All the monkeys were furious about this, so the trainer said, 'All right, all right! I'll give you four in the morning and three in the evening.' The monkeys were delighted. Without having to make any substantive change in the reality of the situation, the monkey keeper was able to manage their pleasure and anger. This is what it means to go along with things."

Of course, *we* are the monkeys the Zhuangzian sage has to deal with. In early China, monkeys served as symbols of willful ignorance and cognitive rigidity, and here they are meant to represent the average person, sure of what he wants and doesn't want, with very full boats indeed. The way to handle monkeys—human or otherwise—is just to let them have their way, if there is no harm in it, rather than insisting on one's original plan. This is "going along with things." We see many of these socially adept exemplars in the text, including a skillful tax collector, who is empty of schemes and responsive to the emotions and needs of those he encounters. This endows him with a powerful *de* that allows him to "collect taxes from morning to night without meeting the slightest resistance." Now *that* seems supernatural!

This monkey trainer story has always reminded me of another story I once heard about a monkey-hunting tribe somewhere in Africa. It's probably apocryphal, but it perfectly illustrates the dangers of getting trapped in one particular way of seeing the world. Supposedly to capture monkeys, a gourd containing food would be staked to the center of a clearing. The opening was designed to be large enough to allow a monkey to reach his hand in but too small for him to withdraw a fistful of food. The story has it that, having allowed the monkey a chance to reach in and grab the food, the tribesmen would then rush out of the bushes to capture him. All the monkey needed to do to escape was let go of the food and run, but—unable to recalibrate his evaluations in light of changing circumstances—he would remain there, panicked and desperate to flee, but with his tightly gripped fistful of food keeping him bound to the gourd. It's the same rigidity that causes an eighty-hour-a-week big-firm lawyer to cling stubbornly to a high-paying, prestigious job even as physical exhaustion, mental breakdown, and an incipient ulcer arrive to take her away.

The poor monkey, and the stressed-out attorney, would do well to emulate the sparrow, the "wisest of birds." According to the *Zhuangzi*, "If its eyes do not spot a suitable place, it will not look twice. If it happens to drop the nut it is carrying, it will simply abandon it and continue on its way. It is wary of people, and yet it lives among them, protected within the altars of grain and soil." The Zhuangzian sage is a bit like this sparrow, living among people, but not getting drawn into attachments that may, in the end, turn into traps. To be clear, the sage likes nuts as much as the next guy. He's just not about to lose his life going after one that's fallen—if it falls, he just lets it go. "The True Person of ancient times slept without dreaming and awoke without worries," Zhuangzi tells us. "He simply ate what was put before him, and his breathing was deep and profound." This is in contrast to "the multitude, who breathe with their throats, oppressed and bent, coughing up their words as if retching," trapped as they are in petty understanding. As Zhuangzi complains at one point:

> Now, as for what most people do and what they find happiness in, I don't know whether, in the end, that is worth calling "happiness" at all. I look at what most people find happiness in—what the masses all flock together to pursue, racing after it as though they can't stop themselves—and I don't really know whether those who say they are happy are really happy or not. In the end, does happiness really exist, or not?
>
> I take *wu-wei* to be the only kind of real happiness. . . . What is

right in the world, and what is wrong, is something that can never
be determined for sure. That being said, let *wu-wei* determine right
and wrong for you! When it comes to attaining ultimate happiness
and invigorating the self, only *wu-wei* can get you close.

This Zhuangzian *wu-wei* is a state of perfect equanimity, flexibility, and responsiveness. Unlike the rigid conscious mind, it can "determine right and wrong" because it doesn't *pre*-determine it. Being
in *wu-wei* is sometimes compared to being like a pivot or a hinge—
the still point at the center from which one can respond to every
change, every eventuality. "When 'this' and 'that' are no longer set
up in opposition, this is called the pivot of the Way," we are told.
"Once the pivot is centered in its socket, it is able to respond inex-
haustibly." Another helpful metaphor is that of letting your mind be
like a mirror:

> Do not serve as an embodier of fame or a storehouse for schemes;
> do not be an initiator of projects or a proprietor of knowledge.
> Fully embody that which cannot be exhausted and wander where
> there are no signs. Use to the fullest what you have received from
> Heaven but do not think that you have gotten anything special.
> Just be empty, that is all. The Perfected Person uses his mind like
> a mirror: he does not lead or welcome; he responds but does not
> store. This is why he is able to win over things and not be harmed.

A literal mirror passively reflects what is in front of it, and when that
thing changes the mirror changes as well. It does not "store" past
images, nor does it anticipate images to come. It simply waits, empty
and receptive. This is what the post-fast consciousness looks like,
open to the world, responding directly from the unconscious mind,
and overflowing with subtle, but powerful, *de*.

At the end of the day, the *Zhuangzi* is a text about individual
wu-wei, about how you, as a person, can learn to move through the

world in a free and easy way. On the surface, you might look like everyone else—going to your job as a butcher, or headed door to door to collect taxes—but on the inside you are quite different because you are now guided by your embodied, hot cognition, not your conscious mind. At one point this ideal is described as being "human on the outside, Heavenly on the inside." It resembles in certain respects the New Testament ideal expressed in John 17, "being in the world but not of it." Like the early Christians, Zhuangzi offers us no concrete political vision. Quite unlike the *Laozi*, he has no interest in transforming Warring States China into a primitive collection of small, isolated villages.

This is arguably because Zhuangzi doesn't believe in fixed prescriptions for how to live. Some, however, have seen this as the primary flaw in the Zhuangzian vision: it's completely selfish, concerned only with individual spiritual perfection and personal happiness. Peter Singer would certainly have sharp words for a follower of Zhuangzi, someone with no plans for dealing with poverty, suffering, political oppression, or social inequity. Zhuangzi seems to accept the political and social status quo, merely giving us a method for moving within it successfully.

There are hints in the text, however, that broader social harmony, and perhaps even societal change, might be achieved if enough individuals were able to enter into *wu-wei*. For one thing, although Zhuangzi tends to emphasize its power to smooth away social friction, his *de* certainly has attractive power. In the second Yan Hui–Confucius dialogue, Yan Hui's success in "forgetting everything"—achieving complete equanimity and freedom from selfishness—inspires even the great Confucius to declare himself ready to become *his* disciple. We see a similar theme in the story of Ai Tuo-tuo, a man described as being "ugly enough to astound the world." Despite his physical appearance, his *de* is so powerful that rulers beg him to be their minister, men fight to be his friend, and women are willing to abandon all hope of marrying just to have a chance of being his concubine. This all because he "harmonized but did not try to take the lead, and focused his awareness only on what was immediately around him."

There are also suggestions in the text that *de* can effect change by having a direct impact on other people's values. If you are sincerely selfless, you can help others to achieve selflessness as well. One story in the *Zhuangzi* concerns a man who has suffered foot amputation as a result of a previous brush with the law—early Chinese penal codes didn't mess around with slaps on the wrist. He describes how he has been changed by having spent time with a particular Daoist master. Before meeting the Master, this former criminal used to go around with a chip on his shoulder, picking fights, angry at the world for his situation. Being in the Master's presence for a short time, though, has completely transformed him, bringing him internal peace and equanimity. "The Master," he declares, "has washed me clean with his goodness."

The idea seems to be that the calm and spiritual equanimity of the Zhuangzian sage is so powerful that it can melt away the spiritual hang-ups of others around him. Simply being in his presence makes you feel better about yourself and more able to enter *wu-wei* on your own. Although Zhuangzi's vision is not overtly political, it presents the possibility of changing the world one person at a time. Where the Laozian ruler's *de* brings everyone in the world into naturalness at once—like a massive, invisible tractor beam—Zhuangzi's sages spread their *de* to others only in one-on-one interactions. Come into contact with them, and you are transformed, and then you go on to transform others. The process seems similar to the "social contagion" effect documented by psychologists, sociologists, and medical health professionals, whereby behaviors or traits—obesity, smoking, excess drinking, depression—appear to spread through social networks, affecting people as far as three degrees of separation away. That is, Joe's depression can produce a noticeable uptick in depression among people he's never even met—people who merely interact with friends of his friends. Zhuangzi's ideal end-state seems to be a world where all people calmly pursue their own naturalness, inter-acting freely with others but always defaulting to a solitary path, like fish "forgetting about each other's presence as they enjoy themselves in the rivers and lakes."

It's an attractive vision. I'll lay my cards on the table and admit that the *Zhuangzi* is, in my opinion, the most profound and beautiful book ever written. In terms of subtlety, insight into the human con-dition, and sheer genius, there is no match in world literature. Only Nietzsche comes close, although his intoxicating brilliance is ulti-mately overshadowed by a darkness and incipient insanity that con-trasts unfavorably with the breezy, healthy optimism of Zhuangzi. That said, Zhuangzi does not ultimately escape from the same ten-sion that plagued our previous thinkers—the paradox of *wu-wei*, how to try not to try—although among all Warring States thinkers he seems the most conscious of it, and the most directly concerned with getting around it.

WHY IS OUR SELF SOMETHING THAT WE NEED TO LOSE?

If there is any logical contradiction lurking in Zhuangzi's thought, we'd expect his friend Huizi to point it out, and Huizi does not dis-appoint. In one of their many dialogues, we find Zhuangzi essen-tially agreeing with Mencius that the tendency to make right and wrong distinctions, and to submit to the rulership of the mind, is part of human nature, and something that distinguishes us from other living things. Zhuangzi refers to this as the "human es-sence," but he doesn't mean that in a good way, as Mencius does. For Zhuangzi, it is our defining flaw, something that needs to be elimi-nated if we are to attain *wu-wei*. As he explains to Huizi: "The sage has the outward physical appearance of a human being but lacks the human essence. Because he looks like a human, he flocks together with other people. Lacking the human essence, though, he does not allow right and wrong to get to him. Lowly! Small! In this way he belongs to the world of humans. Elevated! Great! Standing alone, he perfects his Heavenly qualities."

In this passage Zhuangzi is actually using *essence* as a technical term drawn from Mohist logical theory, one of many he employs. For the Mohists, "essence" is the quality, Y, possessed by a category of things, X, that allows us to distinguish X from all other things in the world. Despite his suspicion of logic, then, Zhuangzi is very well versed in it; his friendship with Huizi may stem from their having trained together at some point. As a Mohist logician himself, Huizi is troubled by the way Zhuangzi is abusing this technical term:

> Huizi said to Zhuangzi, "Can a human really be without the human essence?"
>
> "Yes," Zhuangzi replied.
>
> "But," Huizi countered, "a human without the definitional essence of a human—how can you call him a 'human'?"

This is a perfectly reasonable question; our sympathies lie with Huizi at this point. Zhuangzi tries to dance around it: "What I am referring to as the 'essence' is the tendency to make distinctions of right and wrong. So when I talk about lacking the essence, I am referring to a person who does not let likes and dislikes get in to harm his real self. I'm talking about someone who is able to constantly follow along with the natural and who does not exert conscious effort in his life."

Huizi remains unconvinced:

> "If he doesn't exert conscious effort in his life, how does he manage to get by?"
>
> "The Way gives him his appearance," Zhuangzi replied, "and Heaven gives him a physical form. He just never lets likes and dislikes get in to harm his real self. Now, let's take you! You alienate yourself from your spirit and exhaust your vital essence. You can't

go for a walk without having to lean up against a tree and catch your breath; you can't get through a book without slumping over the desk and falling asleep. Heaven picked out this nice physical form for you, and you use it to twitter pointlessly about petty logical distinctions!"

The dialogue ends here, and we're expected to feel that Zhuangzi has won, with this devastating ad hominem attack pointing out what an unhealthy bookworm Huizi is. The fact remains, though, that Zhuangzi never really answers the question, and that is probably because his plodding, logical friend has gotten uncomfortably close to the basic flaw in the Zhuangzian vision.

According to Zhuangzi, human beings are uniquely burdened with consciousness, language, and explicit values. We are the only species afflicted with cold cognition, which cuts us off from the power of natural, hot cognition happily driving the behavior of all of the other creatures in the world. To join them in harmony with the Way, we need to disengage our cold cognition, eliminating the very thing that distinguishes us from the animals. At the same time, according to the religious framework shared by all thinkers in early China, Heaven made us, and Heaven is, by definition, good. This is why Zhuangzi berates Huizi for wasting this nice physical form Heaven picked out for him, and in other places urges us all to "use to the fullest what you have received from Heaven." But if Heaven gave us our physical bodies, didn't it also give us our internal essence? And if it gave us this internal essence—as it must have, otherwise where would it have come from?—in what sense could it be bad? It's hard to see why Heaven would give us our conscious mind, and our ability to exert "unnatural" effort, if it didn't intend for us to use it. At one point, the *Zhuangzi* tells us that the Way of Heaven is everywhere, even in "piss and shit." If that's so, why is one particular human ability, the ability to use the conscious mind, somehow uniquely cut off from Heaven?

Early Chinese theology aside, this idea that we need to completely eliminate cold cognition in order to be *wu-wei* seems similarly puzzling. As we discussed in some detail in chapter 3, cognitive control evolved for a good reason, enabling us to plan for the future, model hypothetical scenarios, and figure out how to manipulate our environment. Is cold cognition not, therefore, in an important sense "natural" for us? I take this to be the gist of Huizi's complaint about the Zhuangzian sage. People are not fish. They need to think, reason, and exert effort in order to live. This is also the force of a complaint about Zhuangzi that we find in a late chapter of the *Xunzi*, probably written right at the end of the Warring States period: "Zhuangzi was obsessed by the 'Heavenly' and the 'natural,' and so didn't understand the importance of the human." The nature of humans is to be unnatural.

What we are seeing here is basically the same tension we saw in the *Laozi*, and to a certain extent in the *Mencius* as well. We're being told to be natural, but isn't what we're already doing, by definition, what's "natural" for us? The fact that we are not already listening with our *qi* and passively being moved along by our spirits suggests that maybe doing so is not what we're built for. Put in different terms, if the river of the Way is waiting to take us in its current, why are we not already happily floating along in it, lounging on our gourd-rafts, feet dangling in the water, an inner tube full of beer bobbing at our side? When Zhuangzi says to Huizi, in effect, stop worrying about logical issues, stop using your mind to try to figure out right and wrong, isn't he himself guilty of using his mind to set up a "right" in opposition to a "wrong"?

Zhuangzi is aware of this trap and doesn't want to get caught explicitly advocating one way of being over another. The result is some impressive rhetorical dodging and weaving:

> Words are not just wind: words have something to say. But if what they refer to cannot be fixed at all, then do they really say anything? Or do they say nothing?
>
> Now, I am about to say something here. I don't know if, in

saying something, I am just like everyone else or not. Whether what I say is just like what everyone else says or not, it certainly re- sembles what they say in setting up particular categories. In that sense, it is no different from what they say. That said, permit me to say something anyway.

Zhuangzi avoids setting up any specific set of practices that one would have to learn and then consciously put into effect. One imagi- nary dialogue between Confucius and Laozi makes this quite clear: "Confucius said to Laozi, 'Your *de*, Master, matches up with Heaven and Earth, and yet even you must rely upon the perfect teachings of the Way in order to adorn and cultivate your mind. Even among the gentlemen of ancient times, then, who could have avoided such effort?'"

We here see Confucius functioning *as* Confucius for once, not just a mouthpiece for Zhuangzi, and reasonably defending the first *wu-wei* strategy that we discussed: the Confucian project of carv- ing, reshaping, and polishing the self by relying upon the teachings of the past. If Laozi is asking us to behave in a manner that's differ- ent from how we're now behaving, Confucius observes, he must have some practices and teachings to get us there. Moreover, implement- ing them must require effort, at least in the beginning. This view, however, gets definitively smacked down by Laozi: "That is not so! Water has an *wu-wei* relationship to clarity—clarity is simply the natural expression of its innate endowment. The Perfected Person's relationship to *de* is the same: he does not engage in cultivation, and yet things cannot escape his grasp. It is as natural as the height of Heaven, the depth of the Earth, or the brightness of the sun and the moon. What is there to be cultivated?"

This is pretty clear: we Daoists have no truck with *trying*. But of course, that can't be the end of the story, or there would be no Daoism at all. I've noted that we have two separate Yan Hui–Confucius dia- logues in the text, which seem to be two versions of the same story: how uptight Yan Hui forgot about Confucianism and himself and became a Zhuangzian sage. Putting the two versions side by side actually helps us to put our finger on the tension. In the first story, Yan Hui talks to Confucius for a bit, maybe ten minutes at most, and then—bang!—there is no more Hui. He's been suddenly made empty, shocked into *wu-wei*, and can now go off and do his thing. This makes sense if *wu-wei* is our real nature and we just need to realize it. In the second telling, Yan Hui keeps going off and *doing* something—something that takes time, since it's at least a day between each conversation with Confucius. His ability to "sit and forget everything" is acquired gradually, through engaging in a kind of practice, and despite Laozi's protestations in the passage above, this looks a lot like self-cultivation.

My guess is that we have these two versions in the *Zhuangzi* because, among his followers, there was a division into what we could call the sudden versus gradual camps. The sudden approach is more defensible theoretically for a Daoist: being opposed to trying, they don't give you anything to do. They just tell you to wake up. The problem with the sudden approach is that it leaves people in something of an institutional vacuum; a group that gives you nothing concrete to do is not really a religion. On the other hand, if the Zhuangzian school is going to give you something specific to do, this undermines the claim that all trying is bad. The tension that we see here looks very much like the Laozian problem summed up in the claim "He who knows does not speak." Fair enough, but a group of people who claimed to embrace this idea nonetheless felt the need to write an entire book about it. We also hear echoes of Mencius. Just be natural, but no, no, not *that* kind of natural (what you're really inclined to do), *this* kind of natural (the kind that takes some work).

So our tension returns, like a stubborn headache that we just can't seem to shake. We pop painkillers, but they work only temporarily. After an hour or two the throbbing is back, maybe in a slightly different part of our head, but recognizably the same headache. In the medical world, a pain that persists over time and fails to respond to medicine is suspected of being a symptom of a deeper malady. If you've had a headache for two weeks straight, you might want to go in for an MRI. The same is true when it comes to religious or philosophical tensions. In the next chapter we'll try to perform the philosophical equivalent of an MRI, looking below the surface symptoms—a persistent, recurring paradox—to discern the outlines of the underlying disorder. As we'll see, there is a reason that the paradox of *wu-wei* will not go away, and it explains a great deal about the challenges inherent to our civilized, social lives.

7

The Paradox of *Wu-wei*

SPONTANEITY AND TRUST

W<small>E'VE NOW LOOKED AT FOUR DIFFERENT EARLY CHINESE STRAT-</small>egies for attaining *wu-wei* and have seen a consistent tension lurking in the background. All of these thinkers tell us that, if we can just get into a state of complete spontaneity and unselfconsciousness, everything else will work out. We will be in harmony with Heaven. We will possess *de*, a charismatic power that brings social and political success, and we'll move through the physical world with supernatural ease. They all confront us with the problem, though, of how we can consciously *try* to be sincere or effortless. We are being urged to get into a state that, by its very nature, seems unattainable through conscious striving. This is the paradox of *wu-wei*—the problem of how you can try not to try.

This paradox is stubborn, as paradoxes are wont to be. As we've seen, it is no sooner "solved" by one Warring States thinker than it springs up again in a slightly different form. It's like the Hydra of Greek mythology, sprouting two heads for every one that is cut off. In fact, it becomes a central defining issue for *all* East Asian religions. The Chan/Zen Buddhists make a concerted effort to kill it off in the eighth century C.E., when the "gradual" school of Enlightenment (which argued that you needed to work hard at becoming a Buddha; think Confucianism) was officially defeated by the "sudden" school, which declared that we are all Buddhas by nature already, so no effort is necessary (think Daoism, or a more radical version of Mencius). Paradox solved by doctrinal fiat! Anyone calling themself a Chan Buddhist has to advocate sudden enlightenment. No sooner has victory been declared, however, then *another* split develops between what we might call the *gradual* sudden school, which argues that it takes time and work to "wake up" to our true nature, and the *sudden* sudden school, which stomps its feet and says, no, no, we can't do anything *at all* or we're guilty of trying, and we all know that trying is bad.

Japanese Zen inherits the tension, with the fault line running between the Soto versus Rinzai forms of the tradition. Both talk the "sudden" talk—you now *have* to in order to be a Zen Buddhist. Soto, however, gravitates toward what we might call the Confucian end of the spectrum: yes, we are all already Buddhas, but in order to realize that fact we need a whole bunch of rituals and practices, we need to sit *zazen* under the guidance of a teacher, and this is all going to take a long time. Not because we are *trying,* mind you. We do all these practices because this is just what Buddhas do. This infuriates the more Daoist-sounding Rinzai folk, who say, *Kill the Buddha,* burn down the temples, forget *zazen* and ritual bowing and the ringing of pretty little bells and just realize your Buddha nature, *now,* even if that means running around naked and drunk in the mountains howling at the moon. Any involvement in institutions or practices, any type of focused effort, contaminates the true Buddha nature that you need to realize—just wake up. Aha! the Soto folks reply. You Rinzai people are so obsessed with waking up that you don't realize that we are *already* awake, there is no need to realize our Buddha nature, we are already *being* Buddhas, just doing what we are doing, sitting here on a cushion, nothing special. We are, in fact, more sudden than you: you want to *reach* a state of dramatic enlightenment, but we're actually already there. *Boo-ya!*

Contemporary Zen figures continue to struggle with the paradox. For instance, Shunryu Suzuki, a well-known Soto practitioner who taught for many years in San Francisco, hammered home the message to his students that Zen practice is "nothing special." Since we are already Buddhas, we shouldn't get too excited about doing the Buddhist practice. Soto is all about "just sitting," without any expectations or goals. "If our practice is only a means to attain enlightenment," he warns, "there is actually no way to attain it!" The trick is to do Zen practice without being attached to doing it: "Strictly speaking, any effort we make is not good for our practice because it creates waves in our mind. It is impossible, however, to attain absolute calmness of our mind without any effort. We must make some effort, but we must forget ourselves in the effort we make." This could have come right out of the *Zhuangzi*.

The fact that this tension pops up again and again, reemerging at widely separated points in history in more or less the same form, suggests that it reflects a fundamental feature of human life. Indeed, we can see the paradox not only in later East Asian thought—which in any case directly inherited it from Warring States China—but in traditions from around the world. In ancient Greece we find the so-called "Meno problem" in the works of Plato: in order to be taught something, the student needs to recognize it as something worthy of learning. How can you instill love of learning if it does not already exist? Confucius's problem in a nutshell. Aristotle, tasked with training people in the classic Athenian virtues, was forced to conclude that, in order to perform a truly just action, you must already, at some level, *be* just. You have to have the beginnings of justice within you, in which case the role of the teacher is just to help you focus on these stirrings and strengthen them. Sound like Mencius?

This tension continues to pervade later Christian philosophy in the Middle Ages, with thinkers like St. Augustine—committed to the doctrine that human nature is fallen—struggling with what the philosopher Alasdair MacIntyre has called the "Christian version of the paradox of Plato's *Meno*: it seems that only by learning what the texts have to teach can [the reader] come to read those texts aright, but also that only by reading them aright can he or she learn what the texts have to teach." This sounds very much like the versions of the paradox that Confucius, Xunzi, and Mencius were grappling with. Even if we go to India, we can't escape it. The Hindu *Bhagavad Gita* centers on the mystery of *karma yoga*, a state where one can attain the fruits of one's desires only by being sincerely free of desire and nonattached to those fruits. Laozi, anyone?

Moreover, all of this worry about how to attain a state of spiritual spontaneity is, as we saw in the Introduction, intimately related to the purely secular, thoroughly modern problem of why you can't get a date if you want one, can't will yourself out of a slump in sports, and can't win a game of Mindball if you're really trying to win. In fact, once you know what to look for, you start seeing the paradox everywhere. One night I was reading to my five-year-old daughter from her favorite series of books, *Ivy and Bean*, about two elementary school friends who get into various sorts of trouble. In this particular installment, entitled *Bound to Be Bad* (!), Ivy becomes fascinated by the St. Francis of Assisi legend, the take-home lesson of which is apparently that if you are "super-good and pure of heart, animals think you're one of them and they love you and follow you around"—animal *de*! This sounds pretty cool to a seven-year-old girl. So she and Bean resolve to become *really* good so that they can get some of this *de* action and have hummingbirds follow them around. It turns out not to be as easy as they'd thought—*trying* to be sincerely good just doesn't seem to work out for them—so they finally give up and decide to remain bad, which is much more fun anyway. "De . . . paradox of *wu-wei* . . ." I found myself muttering as I read. ("Daddy, *what* are you talking about?" my daughter complained, and bedtime that night was delayed by an extended conversation about what Daddy does for a living.)

So the paradox is real, and it's *everywhere*. The question is why. Why is it so hard for us to relax—really, sincerely relax—if we're not already relaxed, or love a principle that we don't already love, or lose ourselves in an activity that we're not already enjoying? In short, why is *wu-wei* so elusive?

We can move one step closer to solving this puzzle by taking a little ride in the scholarly equivalent of a time machine. Most of the Chinese texts that we have been discussing are so-called "received" texts, or books that have come down to us because, for thousands of years, they have been copied and recopied innumerable times by a series of unknown editors and scribes. We may possess reasonably old—let's say, a couple hundred years old—manuscript versions of them, but it is difficult to say with certainty how the received manuscripts relate to the works penned by the original authors back in the fourth century B.C.E. Another, rarer, type of text is an "archaeological text": an inscription, book, or fragment of a book carved into bone, cast into bronze, or written on bamboo or silk, and then dug out of the ground or discovered in previously sealed-up caves. The exciting thing about such texts is that they have been untouched by copyists or editors since they were buried, and therefore give us a direct window into the conceptual world of ancient China. Two of the most important sets of the archaeological texts yet discovered address the paradox of *wu-wei* in an unusually straightforward fashion and are therefore immensely helpful in pinpointing the origin of the paradox.

STRAIGHT FROM THE EARTH: THE PARADOX OF *WU-WEI* ON BONE AND BAMBOO

Our earliest written records from China come from the Yellow River Valley and have been dated to the second millennium B.C.E. This is smack dab in the middle of the period (roughly 4000–500 B.C.E.) when the first large-scale, literate, agricultural civilizations began to arise in other areas of the world, like Egypt and Mesopotamia. The "oracle bones" were discovered only in the 1890s by a Chinese scholar who noticed that the ox scapula and turtle shells being ground up by local traditional Chinese pharmacists for medicine actually had what looked like writing on them. We don't know how many invaluable records of ancient Chinese life and religion were ground into dust before this scholar came along, but he did launch an effort to decipher the strange script found on these fragments. We now know that these oracle bones, which have since been discovered in vast hoards in ancient tombs, are records of religious divinations performed by the kings of the Shang dynasty (1600–1046 B.C.E.), the earliest of the historically attested Chinese dynasties. They mostly concern queries made by the Shang kings to their supreme god, a figure known as the Lord on High, about famines, wars, sicknesses, and pregnancies in the royal family—all of the issues we typically trouble supernatural beings about.

My own work on the paradox of *wu-wei* was directly inspired by some exploratory essays written by the eminent sinologist and philosopher David Nivison, who wrote back in the 1980s about what he called the "paradox of virtue *(de)*," which he traced all the way back to these oracle bone texts. The paradox of virtue, as defined by Nivison, would look familiar to readers of *Ivy and Bean, Laozi,* or the *Bhagavad Gita: de,* or moral charisma, can be acquired only by someone who is not consciously trying to acquire it. That is, performing a virtuous act with the intention of being rewarded renders your virtue null and void. If you're trying to be good to win a selfish payoff—whether attracting a pretty hummingbird or drawing the entire known world into political order—it's not going to work.

This paradox is evident even in these primordial, cryptic records from China's first great civilization. For instance, Nivison examines one bone inscription that appears to describe an illness that has befallen the consort of the king, and a petition by the king to the spirits to allow him to take the sick woman's place. He expresses his willingness to be afflicted himself if the spirits will shift their wrath away from his beloved. As Nivison explains: "The king's offer of self-sacrifice, ideally, has this result: not only does the sick person get well; further, the king does not himself get sick; and more, because of his willingness to put himself in danger on behalf of another, his *de*, 'virtue,' is magnified." *De*, translated by Nivison as "virtue," seems in these Shang texts to be a somewhat more narrow concept than the one we are working with in the Warring States texts. It refers to a kind of psychic energy that causes other beings—both natural and supernatural—to feel a debt to the possessor, someone like the Shang king, and a consequent desire to obey or help. This is a nice payoff. *But* the whole thing only works if the person involved is not thinking about the payoff at all. The Shang king needs to be genuinely willing to suffer in order to escape suffering. He can get the benefit of having *de* only if he doesn't actually want it.

The oracle bones hail from what was probably the first large-scale, socially stratified, and relatively centralized culture in East Asia. The tension that Nivison identified lurking in these bones is important because it is the symptom of a radical transformation of human life that led to Shang culture and that also occurred independently, around the same time (roughly five thousand years ago), in several places around the world.

For most of the history of our species, we've lived in relatively small, tight-knit bands of hunter-gatherers, interacting primarily with relatives or people well known to us. Evolutionary biologists have very good models for explaining how cooperation works under these conditions. You help relatives because they share your genes, and you cooperate with people you can keep tabs on. I scratch your back, you scratch mine, and if you don't, I'll remember and not scratch yours again. This sort of cooperation is very similar to what we see in other social species. Humans also appear to possess a variety of innate psychological adaptations tailor-made for this kind of small-group lifestyle. These include the ability to recognize and remember a certain number of faces, an eagle eye for social cheating, and emotions—both positive (empathy) and negative (indignation at being treated unfairly)—that seem designed to make social interactions go more smoothly. These are all classic hot-cognition processes. Loving our family, favoring our friends, and getting pissed off when someone cuts in line all come naturally to us.

Possibly the biggest single mystery in evolutionary biology is how one particular primate, us, managed the abrupt transition from our ancient hunter-gatherer lifestyle to the large-scale, urban way of life made possible by agriculture. The innate, spontaneous psychological mechanisms that we evolved over millions of years, and that we share with other primates, appear to be designed primarily to deal with relatives and acquaintances. How were our ancestors, equipped with only these tools, able to adapt to dense urban life, where they were forced to cooperate with large numbers of strangers and take part in novel social institutions, like impersonal, centralized governments? Apes don't pay taxes or recycle. How did we make the leap from our ancient tribal lifestyle to briefcases, law courts, and highrises? Hot cognition by itself was probably not much help, because the time scales involved seem too short for the genetic evolution of new, complex psychological mechanisms. Remember, hot cognition is fast and effective but relatively rigid.

There are basically two theories about how one (relatively) hairless ape managed the transition from tribe to state. The dominant position in the West has been that the transition was made possible because of new *institutions*. According to this model, our hot cognition has not changed at all since we were running around on the African savanna. We are still old-fashioned tribal primates at heart. What *has* changed is the invention of external social institutions— laws and punishments, money, bureaucracies—that allow our tribal instincts to be redirected or repressed. Living in a large-scale society is like performing a perpetual Stroop task: our cognitive control centers are constantly having to override our hot cognition in order to keep us from otherwise natural behavior that, in the civilized world, would get us ostracized or arrested. Civilization is about the triumph of cold cognition over hot. Freud and Mozi would agree.

In recent decades, a growing number of Western philosophers and social scientists have begun to question this view, for reasons that I hope are now obvious to you. Our cold cognition has neither the strength nor the endurance to keep hot cognition in check 24/7. Subjecting people to just a few rounds of the Stroop task, or some other cognitive control exercise, completely exhausts them after fifteen minutes or so; imagine having to do this all day! It is much more likely that the transition to "civilization"—from the Latin *civitas*, or city-state—was managed not by consciously suppressing our tribal emotions but by using cold cognition to extend or redirect instincts through a process of emotional education. We become violently aggressive if someone injures us or one of our relatives; cultural and religious inculcation can train us to react the same way when "our nation"—a group of unrelated strangers I'll probably never meet in person—comes under attack. According to this view, the key to getting lots of strangers to work together is not to create an endless stream of new laws or institutions but to create a set of shared *values*. Laws are something you merely obey. Values are something you *feel*. Once internalized, values function just like other forms of hot cognition—fast, automatic, unconscious, *wu-wei*. Looked at this way, we can begin to see how the paradox of *wu-wei* emerges as a kind of natural consequence of our transition from hunter-gatherers to farmers and city dwellers.

Why this might be so is laid out with admirable succinctness and clarity by Robert Frank, a Cornell economist, and one of the first to reject cold-cognition–only theories in favor of recognizing "the strategic role of the emotions." Most of Frank's colleagues at the time believed that social cooperation was based primarily on rational self-interest, which could be manipulated by tweaking external incentives. If you don't want people to do X (something that people like to do, and that gives them a pleasure yield of, say, 1), then make a law that punishes X with an unpleasant consequence greater than 1. Voilà, you've got social cooperation.

The problem, as Frank observed, is that there are a host of cooperation scenarios—any collaborative projects that require trust, basic economic exchanges, commitments to lovers, spouses, and friends—that simply *cannot* be navigated by rational calculation alone. He argues that *irrationality* is the only way to get cooperation off the ground in these situations and that the precise sort of irrationality we need is provided by human emotions like love, gratitude, indignation, envy, anger, honor, or loyalty.

To understand his point, take a simple economic transaction. Let's say I promise you five chickens in return for working all week on my farm and then give you only three. If you are a purely rational agent, you'll take the three chickens and go on your way. Why? The risk of getting killed or seriously injured in a fight is not worth two chickens. Unfortunately, in acting as a purely rational agent, you'd turn yourself into an economic punching bag. Why would I give you what I promised when I know you'll take less and meekly slink off? Now, imagine that, instead, you are prone to righteous indignation. If I try to shortchange you on the chickens, I know you're going to fly off the handle and do something violent—something *irrationally* violent. A person like you is going to tend to get the number of chickens originally promised.

We no longer tend to denominate our transactions in chickens, but the logic here is identical to the Ultimatum Game we discussed earlier. If I am given $100 to split between you and me, and I offer you $1 and propose to keep $99 for myself, you will get angry and scupper the whole deal, even though the rational thing to do would be to take the $1. Dollars, chickens, whatever: Frank's insight was that even very basic social interactions cannot work unless there are powerful emotions lurking in the background, keeping everyone honest.

In this sense, emotions that seem irrational in the short term—they would result in more short-term costs than payoffs—are rational in the long term. Overall, they reduce costs. However—and here's the rub—they work only if they *really* are irrational and sincere. My righteous indignation is an effective deterrent only if it would actually lead to irrational violence. In other words, the long-term payoff comes only if I don't consciously care about payoffs. Sound familiar?

Frank is keenly interested in this tension and even refers to it as a "simple paradox"—one that is structurally the same as Nivison's paradox of virtue. Our long-term interests require trust between individuals, but trust can be acquired only by those who give up the direct pursuit of self-interest. The conscious pursuit of self-interest is actually incompatible with its attainment. Similarly, in the context of civilization, moral emotions are the key to people obtaining maximum material advantage, but they work only if they are sincere. We couldn't hope for a clearer analysis of the paradox lurking at the heart of any kind of social commitment, whether the Shang king's willingness to take on his wife's sickness or seven-year-old Ivy's sincerity (or lack thereof) in feeling love for all living creatures.

Additional confirmation of an intimate link between the paradox of *wu-wei* and the emergence of civilization is provided by another set of archaeological texts from early China that were discovered in 1993. These texts, written in ink on strips of bamboo and found in a tomb from Guodian (Hubei province) in China, are radically transforming the contours of early Chinese thought. For me, the single most exciting aspect of these texts is that they're deeply concerned with the paradox of *wu-wei* and explicitly place the paradox in the context of social cooperation, centered on a tension between family loyalties and larger loyalties to the state.

To begin with, the Guodian texts clearly embrace something like Frank's model, in which social cooperation works because people share values and fundamentally *trust* one another. These texts emphasize that cultural institutions, such as rituals or laws, will not be effective unless they are deployed by people who are sincerely committed to the public good. In particular, these texts focus on the contrast between two sorts of relationships: the innate father-son relationship and the socially created tie between lord and minister. The former works because it is *wu-wei* by its very nature. Parents naturally love their children, and children naturally love and respect their parents, and they both know that they're stuck with one another no matter what (at least in the ideal world of these thinkers). That means that trust develops organically; as one text puts it, "A father's being treated with filiality, and a son with caring, is not a matter of exerting effort."

Political relationships, on the other hand, are problematic because they are *not* by nature *wu-wei*—the innate hot cognition of a minister does not incline him to trust or obey or love his political superior—but they *need* to be *wu-wei* in order to work properly. The Guodian corpus is permeated by an anxiety about how to make this transition; while one cannot consciously try to be moral, one also cannot *not* try, since political order depends on developing public virtues in the aspiring official. After baldly declaring that benevolence and rightness cannot be achieved through striving, one of the texts then revealingly continues: "If you try to be filial, this is not true filiality; if you try to be obedient, this is not true obedience. You cannot try, but you also cannot *not* try; trying is wrong, but not trying is also wrong." The overall tenor of the Guodian corpus is that, though striving or trying is morally suspect, one cannot help but try if the world is to be properly ordered and human beings are to escape from the chaos of the state of nature. Therefore, if one is to order society, it is necessary to find a way around this paradox—the paradox of how you can try not to try.

One of the nice things about these recently discovered texts is that they make it clear how the paradox of *wu-wei* is inevitably tied up with the value-based model of cooperation. If you're just using rewards and punishments—the rational, self-interested, cold-cognition strategy—it doesn't matter what people *feel* on the inside. You set up the incentives, let people figure them out, and then judge them purely on their behavior. In the values model, on the other hand, what people are really feeling on the inside is crucial: if I can't trust that you're committed to the same ideals that I'm committed to, there's no way we can work together.

TATTOOS AND SHIBBOLETHS:
IN THE BODY WE TRUST

The more clever, or sinister, among you must by now be thinking, fair enough, if I'm going to be successful in this world, I need to be viewed as someone committed to the greater good. Wouldn't the best strategy, then, be to merely *seem* committed, and therefore reap all of the benefits of long-term commitment, while in fact remaining at heart a rational, self-serving bastard, willing to abandon my side of the bargain anytime it proves too costly for me?

The answer is, absolutely yes. In fact, in any given population there will be a group of people pursuing precisely this strategy. One of the more robust findings among people who like to mathematically model cooperation strategies is that the commitment strategy scenario will inevitably result in a stable mix of "cooperators" (truly committed individuals) and "defectors" (wolves in cooperators' clothing), with a spectrum of more or less cooperative people in between. A community consisting entirely of defectors would never get off the ground. A community consisting entirely of cooperators would very quickly get invaded by defectors enjoying all of the benefits and paying none of the costs and would then collapse under the weight of these freeloaders. The situation that has come to prevail in large-scale human societies around the world is one in which cooperation is quite prevalent but some defectors can survive because the cost of perfect cheater detection is too high. Pervasive suspicion is as paralyzing as blind trust is open to abuse.

For this balance to work, though, defectors need to be kept to a manageable number. For *this* to happen, in turn, cooperators need a relatively reliable signal that they can use to quickly and efficiently identify one another, so as to preferentially cooperate with fellow good guys and shun contact with bad guys. For this signal to be effective, it has to be relatively difficult for defectors to mimic. You need good methods for both policing group boundaries—making sure that in-group members are clearly distinguished from out-group members—and figuring out which of your fellow in-group members are real cooperators and not selfish free riders.

One possible solution is to focus on signals of honesty that are relatively hard, if not impossible, to fake. This idea of signaling arose when biologists tried to figure out some puzzling bits of animal behavior. Peacocks, for instance, have ridiculously large, colorful tails. These tails are expensive to produce, constitute a huge neon "eat me" sign on the animals they're attached to, and also make it very difficult for the animals in question to flee when some predator answers the invitation. Gazelles, to take another example, do not immediately run off when they are approached on the savanna by a predator. Instead, they first waste a bunch of energy and time by "stotting"— leaping up and down in place, as if taunting the predator and inviting it to come after them.

It's now thought that the most likely explanation for these otherwise puzzling phenomena is that huge tails and wasteful hopping serve as reliable signals. A peacock with an enormous tail is broadcasting to females the message, "Hey, look, I'm so healthy and strong that I can produce this ridiculous tail and somehow still survive while dragging it around. Mate with me, I'll make strong kids." The stotting gazelle is sending a signal like the one broadcast by a pimply teenager revving his souped-up car at a stoplight: "I've got a big engine, don't try to race me."

The key feature of these biological signals is that they are *inherently* unfakeable, because of physical limitations. A sick peacock simply couldn't produce or survive with a big, colorful tail, and a weak gazelle just can't leap very high into the air. Many anthropologists believe that these types of inherently unfakeable signals are also used by people to broadcast social reliability and loyal membership in groups. This would be a case of cultural, rather than genetic, signaling.

A good candidate for such cultural signaling is the widespread practice of bodily alteration or scarification. In many societies a common way to signal commitment to a group, and its values, is to physically alter your body in a permanent, and ideally painful, way. Cut off the foreskin of your penis, tattoo your face, radically elongate your earlobes: it's hard to go back on these choices. This is something I tried to impress upon the teenage son of a colleague when he embarked upon a campaign to cover his entire body in somewhat obscene, anticapitalist tattoos. "You're only nineteen," I would say to him. "What if you decide, at age thirty-five or so, that you'd rather not have a tattoo of a fat capitalist squeezing blood out of the globe on your right forearm?" His reply was very revealing: "I don't ever want to be the kind of person who wouldn't want that tattoo." This cuts right to the heart of costly signaling. It's a sign of intense and sincere commitment to a certain set of values precisely because it's *permanent*.

Another somewhat obvious and hard-to-fake sign of group membership is *accent*. After about age twelve or so, human beings begin to lose the ability to acquire a new language perfectly, which means that any foreign language learned as an adult will be spoken with a noticeable accent. This is probably just a by-product of the neural pruning that goes on in human development, because in a normal evolutionary environment we've learned all the languages we're going to need by age eleven, so expensive language-learning neurons can be absorbed and put to better use elsewhere. It seems, though, that cultural evolution has zeroed in on this feature of human physiology as a reliable signal. If you didn't grow up with us, you can't speak like us, and we're going to be less likely to cooperate with you. This was famously used to great effect by the men of Gilead in the biblical Book of Judges (12:4–6). Faced with the problem of singling out the defeated Ephraimites for slaughter, they demanded that anyone crossing the river to flee "say *shibboleth*." Apparently the dialect of the Ephraimites lacked the retroflex *sh-* sound, so the best they could do was *sibboleth*, at which point they'd immediately be killed. The word *shibboleth* in English now refers to an arbitrary sign of being part of an in-group, but that's not at all its original meaning: the reason the biblical *shibboleth* worked was that it was anything but arbitrary. All groups use their own versions of the shibboleth in situations where defining membership is crucial.

Obvious group markers like tattoos and accents help to stabilize social groups by clearly distinguishing between us and them, creating barriers to switching sides if the going gets rough, and functioning as reliable signals that a stranger is likely to be trustworthy. This alone goes a long way toward solving some of the cooperation problems inherent to the value commitment model. It's likely that this is why successful religions, for instance, typically require their adherents to spend enormous amounts of time acquiring arcane knowledge (learning dead languages, memorizing scriptures) or participating in otherwise apparently pointless behavior (sitting in church every Sunday listening to boring sermons instead of farming or trading). One's willingness to put in the time and effort sends a strong signal of commitment to the group.

Even within such groups, however, you've still got the problem of distinguishing gung-ho, truly committed cooperators from others who are just along for the ride. You need to be able to discern genuine *commitment* from mere enlightened self-interest. It's possible, for instance, that I take communion and go to church every Sunday because I know that, in this community, being part of the Christian club brings awesome payoffs: social support networks, preferential access to loans and contracts, better schools for my children. In this case, it might just be worth my while to learn the Bible and show up regularly for services if the long-term payoffs outweigh the costs. Therefore, although some theorists of religion use the terminology of "signaling" to refer to all group-marking behaviors, there isn't a perfect analogy between church attendance and peacock tails. A peacock with an impressive tail is physically healthy, end of story. A signal like religious observance, however, *could* be faked, in the sense that the signal it sends (I really care about God and share the values of this community) may not accurately reflect the real underlying motivation (I'm only here to network over free muffins and coffee).

This is where more subtle physiological signs become important.

The conscious mind can be a lying, conniving manipulator, but one of the salient features of hot cognition is that it's typically *not* subject to conscious control. Our bodies tend to tell the truth. All of us have had the experience of being betrayed by an annoying body that won't go along with an excellent lie that our minds have cooked up. When I am lying—or even *thinking about* lying—I develop an odd smile. I know only too well what it feels like from the inside: as I start to lie, I can feel the muscles in my cheeks begin to contract against my will, and it only gets worse if I try to repress it. I never knew what it looked like from the outside, though, until the day I saw it reproduced exactly in my daughter. When I suspect that she is lying about something, I make her repeat herself while looking straight into my eyes. If she's telling the truth, she can keep a straight face, but if she's lying, I can see the awkward, unnatural smile forming around her mouth, and if I call her on it and she tries to control it, it becomes even more pronounced. The goofy "lying smile" is apparently one of many genetic burdens my daughter has inherited from me. She'll probably get better at lying with age, but right now, while she's still five, I've got a more or less 100 percent effective lie detector at my disposal.

The psychologist Paul Ekman has become famous for his work on cataloguing and explaining the physiological basis for hundreds of human emotional microexpressions, as well as for his cultivated ability to identify these expressions in the real world. One of Ekman's most important findings is that sincere emotional expressions tend to be executed by muscle systems that are very difficult, if not impossible, to bring under conscious control. We are all familiar with fake smiles from family photographs. These awkward smiles can be clearly distinguished from what Ekman calls the "Duchene smile," named after a nineteenth-century French neurologist who demonstrated that sincere smiles—smiles instigated by our hot, automatic systems—use an entirely different set of muscles than contrived smiles. The reason that urging your photographic subject to "say cheese" sometimes works is because the request is so absurd that it results in genuine amusement, which then causes a small, sincere smile to appear. Emotional reactions serve as reliable signals precisely because they are not typically under conscious control.

There are, of course, always going to be outliers who are very good at faking. Robert Frank illustrates this point with a great stock photo of the director-actor Woody Allen looking at a camera with that classic, Woody-Allenesque expression of resigned sadness, which is crucially dependent on the ability to use one's frontal, corrugator, and pyramidal muscles to elevate the eyebrows at the center of the brow. This is the expression that all of us make spontaneously when we experience sadness or distress, but most of the population cannot produce it voluntarily like Allen can. If everyone, or even a large minority of people, could easily control these muscles, they would cease to serve as useful signals.

The best way to understand Woody Allen's ability is that he, and people like him, represent a small group that has gained a temporary lead in the evolutionary arms race between cheaters and cheater detectors. The better our particular line of primates has become at faking emotional signals, the better we've become at detecting such faking, so we naturally see a spectrum of abilities. It's likely that professional fortune-tellers or astrologers, for instance, are essentially anti–Woody Allens: unusually good at reading body language and subtle behavioral clues, which in turn helps them deliver astoundingly accurate appraisals of strangers. They are as good at reading the real emotions behind subtle facial expressions as Woody Allen is at faking those emotions. My ability to tell when my five-year-old daughter is lying—or my wife's ability to tell when I am lying—would seem just as magical and astounding to a chimpanzee. After another thousand years or so of evolution we'll probably *all* be as good as Woody Allen at looking mopey and morose, but endearingly and intelligently so. Once this happens, our species will have to move on to new—and harder-to-fake—signals to pick out sincere, well-intentioned, but misunderstood intellectuals.

In the search for reliable signals of people's genuine commitments, we also use other physical features—general appearance, body language, gait—to make rapid, and surprisingly accurate, judgments about whether a stranger is likely to be a good cooperator. There is now a huge, and continuously growing, literature on "thin-slicing": our ability to assess people based on extremely small samples of behavior (on the order of a couple seconds) or even still photographs. We can determine, almost instantly, others' personality traits, job performance abilities, sexual orientation, and propensity to cheat or engage in violence. The clues employed seem to be a combination of dynamic microexpressions (including posture and gait) and more constant aspects of one's appearance, like the width of the face or size of the chin.

Thin-slicing accuracy seems better than chance under normal conditions, but not perfect. Some people are much better at it than others, and one's ability can be improved by training, but there's constant pressure to weed out fakers. To enhance our natural thin-slicing abilities, humans have therefore also developed various cultural practices that make these instant assessments more reliable. These techniques take advantage of the fact that deception is fundamentally a cold-cognition act and relies on cognitive control centers. This means that if we can impair the cognitive control abilities of people we're trying to judge, we'll do a better job of sussing them out: they will be less able to confuse our cheater-detection systems.

In one study that has proven enormously useful to law enforcement agencies, researchers found that police officers could significantly improve their ability to detect false statements if suspects were asked to give their alibis in *reverse* order, starting with the most recent event and working their way back. This is not the way we normally tell stories, so being forced to do it increases cognitive load. Dishonest suspects, it turns out, are less effective liars if you handicap their conscious minds in this way.

This reverse-order alibi technique is a great tool for law enforcement but not terribly practical when evaluating a potential business partner or deciding if the people you're about to make a peace treaty with are being sincere. There are other ways to achieve the same effect, though. The police study aimed to reduce subjects' cognitive control ability by increasing the load—adding more weight, as it were. Alternately, you can keep the load constant but decrease cognitive control ability—weaken the mental muscles—by suppressing cognitive control centers. One way to do this is transcranial magnetic stimulation (TMS), which involves applying a powerful magnetic force to the appropriate region of the skull. TMS, however, is a very recent technology and not exactly widely available. Also, in most cultures it's considered bad form to shock the heads of new acquaintances with huge magnets.

A much more low-tech and socially acceptable way to produce the same effect is to get someone completely *wasted*. As we discussed in chapter 6, one of the primary effects of alcohol and other intoxicants is to "downregulate," or temporarily paralyze, areas of the prefrontal cortex associated with cognitive control. A couple shots of tequila is the liquid equivalent of a nice jolt of TMS. It's therefore no accident that intoxicants of various sorts are frequently employed by human beings as social lubricants. Alcohol, kava, cannabis, magic mushrooms, you name it: any intoxicant that people can get their hands on quickly comes to play a central role in social occasions, both formal and informal. In ancient China, no major treaty was signed without first bringing everyone together in an extended, alcohol-soaked banquet. In fact, this is one feature of Chinese culture that has not changed a bit in over four thousand years. Any modern businessperson hoping to ink a deal with Chinese partners had better get his or her liver in shape first.

On a less formal level, this is no doubt why intoxicants are a universal feature of all sorts of human social gatherings, from casual cocktail parties to fraternity mixers. Not only is getting drunk pleasant, it also typically causes people to get along more freely and easily (at least to a certain point, after which the drunken fights break out). Intoxication enhances cooperation in at least two ways. First of all, it reduces social faking by inhibiting cognitive control centers. Second, if we all get drunk together, we create a situation of mutual vulnerability that makes trust easier to establish. Getting drunk is essentially an act of mental disarmament. In the same way that shaking right hands with someone assures them that you're not holding a weapon, downing a few tequila shots is like checking your prefrontal cortex at the door. See? No cognitive control. You can trust me.

Similarly, one of the reasons that emotions are useful social signals is that they utilize physiological pathways that are *relatively* beyond conscious control. Few pathways in humans, however, are entirely beyond the reach of the cognitive control regions: even deep-seated and relatively automatic emotions can be brought under conscious control with practice and training. As the case of Woody Allen illustrates, people who *seem* sad and well intentioned sometimes are not. Emotions are imperfect signals because some people are incredibly good at controlling them consciously, and even those who are not particularly gifted in this regard can sometimes manipulate the otherwise honest signals being kicked off by the unconscious mind.

Importantly, though, this conscious effort *itself* throws off its own signals. You don't necessarily need an fMRI machine to know that someone's up to something fishy. Most people are familiar with the obvious signs of dishonesty: sweating, rapid side-to-side eye movements, difficulty making direct eye contact. There are, however, also more subtle ones. For instance, a large experimental literature has shown that, when you exert mental effort, the pupils of your eyes become slightly dilated. This is probably why we try to look into the eyes of people we're not sure we trust, and why untrustworthy people avoid eye contact. At some level, their hot-cognition system is saying, "Don't let them see how big your pupils are!"

The role of cognitive control in dishonesty was nicely illustrated in a recent brain imaging study that subjects were (falsely!) led to believe was about their ability to predict the future. The experimenters had subjects in fMRI machines predict the outcome of a series of computerized coin flips that they watched on a video screen. They were asked to make a prediction, heads or tails. Then a screen flashed up reading "Heads" or "Tails," and the subjects were asked if they had predicted the outcome correctly. If they had, they were rewarded with money. In one of the conditions, the subjects were required to publicly record their prediction, which left no scope for cheating. In the other (more interesting) condition, they were merely asked to privately remember it: when cued with the question "Did you predict correctly?" they had a financial incentive to lie. If they clicked no, they didn't get the money. The beauty of this design is that, with enough trials, a clear signature of deception emerges—being correct at a rate significantly greater than chance—that allows you to identify who has been lying, and how much he or she has been lying, although you cannot pick out individual instances of lying, since only the subject actually knows what he or she predicted. What the experimenters found was a pattern of activity in the cognitive control regions of dishonest subjects—both when they must have been lying and when they were consciously *refraining* from lying—that was completely absent in the honest subjects.

Similar studies have suggested that—at least for inhabitants of Western, industrialized cultures—people are spontaneously generous if forced to make instant decisions but begin to gravitate toward more selfish strategies if given time to think. All of this suggests that honest behavior is governed by automatic mental processes, whereas controlled processes are involved in lying or faking. In other words, effortless, unselfconscious behavior—behavior that is *wu-wei*—acts like a window into our true character.

IT'S A REAL PARADOX: *WU-WEI* AND *DE*

We've seen that the early Chinese thought that people in *wu-wei* possess *de*, "charismatic power" or "virtue." We've also seen that, for all of our thinkers, *de* is a crucial factor in why people in *wu-wei* are successful in life. For the Confucians, it is the power that a trusted ruler relies upon to attract loyal ministers or that such ministers use to signal their loyalty. For the Daoists, *de* announces to others that you are being spontaneous, and this both puts them at ease and helps them to relax into naturalness. For both Confucians and Daoists, you acquire *de* only if you are sincerely committed to the Way—the power of *de* serves as a palpable, unfakeable signal of commitment to the group's values. While the physical efficacy that comes with *wu-wei* is seen as important, at the end of the day it's the social effects that are the most valued.

Looked at from a contemporary perspective, we can redescribe *de* as the body language that someone exhibits when their cognitive control centers are downregulated—when they are being genuinely spontaneous. Confucius advises that, when judging the character of someone, you should pay attention to their eyes, their offhand remarks, their subtle body language, especially when they think that no one else is around. If you do that, he says, "how can a person's true character hide?"

In other words, someone who is truly committed to the values of his society has completely downloaded them into his embodied mind. Proof that this internalization process is complete can then be accurately read off their body, in a moment of thin-slicing. Laozi's sages have faces as smooth and untroubled as infants, and the Zhuangzian sage's free and easy nature is reflected in every feature:

> The True Man of ancient times slept without dreaming and woke without a care; he ate without greed and his breath came from deep inside. . . . His face was calm, his forehead broad. . . . His bearing was elegant but not rigid; he appeared to lack but accepted no gifts; he was relaxed in his solitude but not pretentious, generous of spirit but not ostentatious. Mild and cheerful, he seemed happy; calm, he went along with what could not be avoided; annoyed, he let it show in his face; relaxed, he rested in his *de*. . . . Restrained, he seemed to prefer his own company; bemused, he forgot what he was going to say.

Who wouldn't want to hang out with a guy like this? And who would not immediately trust him?

De is powerful, then, because it reveals who you really are. Not who your conscious mind thinks you should be in this particular moment, for this particular audience, but how you really are when you have relaxed into hot cognition. There is a wonderful passage from an early Confucian text called *The Great Learning* that focuses on the practice of being "vigilant while alone." This practice is predicated on the belief that an aspiring Confucian gentleman can get an accurate sense of how close he is to attaining true *wu-wei* by analyzing his own behavior and expressions in his most unguarded, private moments:

> The traditional saying "Make your thoughts sincere" refers to not indulging in any self-deception. You should love the Way without thinking, the way that you hate a disgusting smell or love a beautiful object. This is called spontaneous enjoyment. This is why the gentleman must be vigilant while alone! When the petty person relaxes at the end of the day, the vices in which he will indulge know no limits. When confronted with a true gentleman, though, the petty man is then instantly anxious and tries to conceal his vices, putting on a show of being good. No matter how much he tries to deceive, however, it is of no use: when others look at him, it is as if they can see straight into his heart and liver—what use is trying to hide! This is what we mean by saying that inward sincerity can always be seen from the outside. . . . Just as riches adorn a mansion, so does *de* adorn one's person. The mind is relaxed, the body at ease. This is why the gentleman must be vigilant while alone.

Most of us have experienced the uncomfortable feeling of having another person "look straight into our heart and liver." If we're trying to be something that we're not, it usually shows. *Wu-wei* reveals your inner character—your *de* or lack of *de*—not only because it's automatic, and thus not subject to the conscious spin-doctor, but because the very fact that you're not exerting cognitive control indicates you have no need to. You're like the honest subjects in the coin-flip study: it doesn't even occur to you to cheat, so you don't need to stop yourself from cheating. Self-confidence sends the signal not only that you're happy—you are engaged in activities that are genuinely pleasing to you—but that you are what you claim to be. Relaxation and absorption in something that is valued—true *wu-wei*—is thus a sign of genuine commitment to the activity and its larger framework. If you're not enjoying singing the hymns, maybe you're not a real Christian. Maybe you're just *pretending* to be a Christian in order to get the benefits of being part of the group. So let me watch you when you're singing and you think that no one is looking.

The connection between *wu-wei* and *de* thus makes perfect sense from an evolutionary perspective. *De* is the attractive vibe—a combination of body language, microemotions, tone of voice, general appearance—kicked off by people who are honest, sincere, self-confident, and relaxed. It's attractive because it's a relatively hard-to-fake signal of a trustworthy cooperator, and the logic of civilized life makes us very keen to distinguish reliable cooperators from unreliable defectors. And the best time to look for these signals of reliability is when everyone's guard is down: when we're dancing, singing, drinking, and playing. As we saw back in chapter 2, a key feature of *wu-wei* is the sense of being absorbed into a larger, valued whole, whether that involves the joys of being with a particular group of friends gathered around a particular kitchen table, or with a certain congregation, or surrounded by the beauty of a particular landscape. The lack of *wu-wei*—and consequent lack of *de*—therefore serves as a reliable indicator that I don't care, I do not feel myself effortlessly absorbed into our conversation or our religious ceremony.

We are now in a position to see why the tension of how you can try not to try is not just an accident of a particular philosophical or religious tradition—a weird quirk found only in China or East Asia—but rather a structural feature of civilized life. The key is to see that, at the end of the day, it is not enough that a trusted member of our group is able to dance our dances, drink wine with us, and speak our language well, although that certainly doesn't hurt. Indeed, as we've seen, such relatively unfakeable markers are probably very useful for quick *Is he one of us?* assessments. At the end of the day, though, we need more. Because of the constant danger of free riders faking commitment—putting in the time to learn our cultural skills, while secretly ready to betray our shared values as soon as an opportunity presents itself—a potential gap opens between external behavior and internal motivation. What we want, then, is a particular type of desirable, hot behavior where there is absolutely no gap between action and motivation. We want to assure ourselves that there is no extraneous cold cognition sneaking around backstage with potentially nefarious plans of its own. What we're interested in is not mere physical skills but what philosophers call "virtues": stable dispositions to perform socially desirable actions in a manner that's sincerely motivated by shared values.

A simple example will suffice to explain the difference between these two types of hot cognition, virtues and skills. A skill, like piano playing, certainly involves hot cognition: you could never get beyond painful, slow renditions of "Mary Had a Little Lamb" without relegating much of your action to relatively automatic systems. The value of your piano playing, however, is ultimately dependent only on what you produce when you're sitting there at the keyboard. Internal motivation is, in the final analysis, irrelevant. Having downloaded most of the execution of a given musical piece into your basal ganglia and sensorimotor systems, you are free to do whatever you want with the cold-cognition abilities you have left over. Although we might romantically suppose that the performer of a profoundly moving and beautifully executed piano sonata is completely absorbed in the music she is playing, experiencing the same deep emotions that the piece inspires in us, we could hardly fault her—or ask for our money back—if we subsequently discovered that she had, in fact, been mostly thinking about what she was going to have for dinner that evening. The performance stands on its own merits, regardless of the internal state of the performer. Things are rather different when it comes to virtues, though, because virtues are fundamentally about social cooperation, which is inherently vulnerable to free riding. Virtues have to be sincere to count as real virtues: there can be no daylight between external act and internal intention.

On November 14, 2012, a tourist in Times Square surreptitiously snapped a picture of a police officer kneeling down to help a barefoot homeless man put on a new pair of boots. When posted to the NYPD's Facebook page, the photo went completely viral. The officer, named Lawrence DePrimo, had apparently been so moved by the suffering of the barefoot man that he popped into a nearby shoe store to buy him a new pair of boots with his own money. "It was freezing out and you could see the blisters on the man's feet," he said when asked about the incident. "I had two pairs of socks and I was still cold." The story was an enormous publicity coup for the NYPD, but the secret to its appeal was the spontaneity of the officer's gesture and the happenstance of someone catching it on film. Imagine if we found out later that DePrimo knew the photographer was there and had been merely grandstanding for the camera—his act motivated by desire for fame rather than spontaneous compassion. This knowledge would instantly transform a seemingly heartwarming act of kindness into a horrible travesty. The very act itself would magically change, even though nothing would be materially different: the officer would still be out $75, and the homeless guy would still have a nice pair of boots that he didn't have before. We have a powerful, ineradicable intuition that a "compassionate" action performed without the right motivation is merely a semblance, a counterfeit of virtue. The flip side is that evidence of sincerity and spontaneity in the moral realm inspires and moves us.

Interestingly, the ancient Greek philosopher Aristotle was very much aware of this issue, which he framed in terms of a distinction between physical skills (what he called "crafts") and virtues. When it comes to a skill like playing the piano, we have no problem imagining how forced training can eventually result in an internalized, reliable skill. Although many kids who are forced to take years of piano lessons end up hating the piano, for a lucky few this grind in the end produces both a high degree of talent and genuine pleasure as well. When it comes to a virtue like compassion, however, it's not at all clear how training alone could work. We have an intuition that, if I don't spontaneously feel at least some compassion for the homeless guy with the blistered feet, going through the motions of buying him new boots is not going to transform me into a compassionate person. Even if I force myself to act out compassionate behaviors *hundreds* of times, as if practicing scales, it's hard to see how or when I would thereby become a genuinely compassionate person. There's an important sense in which character virtues strike us as the kind of thing that can't be acquired just by trying: either you have them or you don't.

It is this difference between skills and virtues that, at the end of the day, is driving the paradox of *wu-wei*. Although our Chinese thinkers used stories like that of Butcher Ding or Woodcarver Qing to illustrate aspects of *wu-wei*—as did Aristotle, incidentally—the kind of *wu-wei* that they are worried about is *moral* in nature. They want to foster cooperation and virtue, not efficient butchery. So the paradox exists because the kinds of virtues that people care about and value in others center on *who you are*, not necessarily what you do. They are about stable, *inner* states, not just outward behavior. They are about *values*, not merely actions, because it's commitment to shared values that allows large-scale societies to function. So it's not enough to perform generous actions, you need to become a generous *person*. This is an enormously difficult trick to pull off, which is why true *wu-wei* is both inherently hard to reach *and* such a great signal of trustworthiness once we've managed to get there. We're attracted to genuinely *wu-wei* people—they have *de*—because evolution has shaped us to home in on signals of sincerity that are difficult to consciously simulate and even harder to experience on demand, and to do so in response to basic challenges inherent to human cooperation.

Where does this leave us? If the paradox is a real one, arising from basic structural features of human civilized life, it is unsurprising that none of our Chinese thinkers was able to come up with a single, surefire solution to it, and that people in widely separated cultures have struggled with the same tension. Indeed, that's why we call it a "paradox"—if there were an answer, we'd demote it to a mere "problem" or "puzzle." Paradoxes are not something that you *solve*, they are something that you learn how to live with. How we might go about doing so is the topic of our next, and final, chapter.

8

Learning from *Wu-wei*

LIVING WITH PARADOX

WHAT IS THE ANSWER FOR THOSE OF US WHO WANT TO BE *WU-wei* but are not? In our journey through early Chinese thought, we seem to have ended up back where we started. This circularity characterizes later East Asian religious thinking, where "trying" and "stop trying" strategies succeed one another without either decisively coming out on top. The tension also pervades other world religious and philosophical traditions. On the face of it, this seems odd: you'd suppose that religious thinkers, over the course of a couple thousand years, would gradually manage to figure things out. But when Shun-ryu Suzuki tells his American disciples in twentieth-century San Francisco that trying is bad yet they need to try anyway, he could be directly paraphrasing from 2,300-year-old muddy strips of bamboo pulled out of the earth at Guodian: "You cannot try, but you also cannot *not* try; trying is wrong, but not trying is also wrong." That's a bit eerie. Are we no better off today than the person who penned these characters on bamboo in the fourth century B.C.E.?

We are not, but at least we now have a fairly good sense *why* not. The paradox of *wu-wei* arises out of problems surrounding human cooperation and trust, and its paradoxical nature is not ac-cidental but rather a design feature. If it could be easily conjured away through some doctrinal innovation or new self-cultivation technique, it wouldn't be doing its job. At the same time, it must be possible to skirt the paradox *in practice*—there has to be some way for a person who is not in *wu-wei* to somehow get there. Otherwise we'd have no effortlessly virtuous Confucius, no genuinely desireless Laozian sage. Insomniacs would never fall asleep, self-centered children would never learn real concern for others, and desperate singles would never get a date. Indeed, as we saw in the last chapter, the whole project of civilization would fail.

Fortunately for us, the early Chinese explored every conceivable strategy for moving a person from a state of alienated trying into perfected *wu-wei*. You can carve and polish: subject yourself to rigorous, long-term training designed to eventually instill the right dispositions. You can embrace simplicity: actively reject the pursuit of goals, in the hope that the goals will then be obtained by themselves. You can cultivate your sprouts: try to identify incipient tendencies of desirable behavior within you, and then nurture and expand them until they are strong enough to take over. Or you can just go with the flow: forget about trying, forget about not trying, and just let the values that you want to embrace pick you up and carry you along.

Which of these strategies is best? More to the point for us in the modern world, what *can* we do if we can't get a date? How can we try to meet someone without creating the anti-*de* that acts like a repellent force field? If we're Steve Blass, watching our life's dream disappear before our eyes because we can't just relax and love throwing a baseball again, what are our options? What if we're not as compassionate or wise or courageous or understanding as we'd like to be? Maybe we're happy just the way we are, but most likely we're not. In any case, any religion I've ever studied is motivated by the sense that there's at least something we need to change about ourselves or our relationships. The problem is how to consciously strive to do this without fatally blocking *wu-wei*.

After almost 2,500 years of serious, worldwide effort, no one has come up with a completely foolproof solution to this problem. This is both because the paradox is a genuine one and because the different strategies we've explored vary in their appropriateness in at least two ways.

To begin with, different strategies may simply suit different types of people. Those with conservative personalities—who are often, though not always, politically conservative as well—generally take a dim view of human nature and emphasize the importance of tradition, authority, and discipline. Liberals tend to have a sunnier view of human nature and therefore place more stock in individual autonomy, creativity, and flexibility. Looked at this way, the swings between Confucian and Daoist strategies for attaining *wu-wei* could be seen as an alternation between conservative and liberal responses, with Mencius trying to take a moderate view. The same could be said of the "gradual" versus "sudden" debate in Zen Buddhism.

There is considerable evidence that a basic inclination toward either liberalism or conservatism is a heritable trait; like other personality traits such as extroversion or introversion, it has a partially genetic basis. Just as people come into the world being relatively open or closed to new experiences, extremely conscientious or fairly loosey-goosey, it seems that people are also born with liberal or conservative inclinations. So it may be the case that whether you find yourself drawn to the carving-and-polishing or letting-go strategies has something to do with where you fall along this spectrum. Seeing the different strategies as having at least a partial basis in innate personality differences also goes a long way toward explaining why no single strategy ever becomes dominant for any period of time: as soon as one strategy gets established as orthodoxy, the opposite strategy is quickly reasserted by those who have different inclinations. This would make sense if human populations consist of a mix of liberal- and conservative-leaning people, each inclined to push back against strategies that go against their own innate grain.

Different phases of life might also call for different strategies. Carving and polishing might be more appropriate earlier in life, or when you are just learning something new. There is good evidence that, when it comes to skill acquisition, conscious attention to technique and explicit feedback is actually very helpful. It's only when you reach the expert stage that cold cognition begins to disrupt your performance. The same may be true of morality. A deeply ingrained moral disposition could become too rigid as you age, in which case you might need to shift to the sprout or letting-go approach. It's often said that in traditional China people were Confucians when employed as officials but Daoists when they got axed by a new regime or court faction and sent off to the countryside. Which strategy helps to maintain you in *wu-wei* may very well change over time in response to different career or family demands.

Temperament and age aside, it's also clear that different situations will demand different approaches to the problem of failed *wu-wei*. This is as true for societies as it is for individuals. For instance, some believe that our present cultural moment calls for a healthy dose of Confucianism. Perceptive thinkers throughout history have seen the Confucian strategy of creating a new, artificial nature as crucial for civilized life. In the modern West, we've come to view social rituals as constraining, alienating, or inevitably hypocritical, but we might want to give Confucianism a bit more credit on this front. Indeed, the continuing enthusiasm for Zen Buddhism and Daoism that began sweeping through Western popular culture in the 1960s is gradually being supplemented by voices arguing that we have something to learn from stuffy Confucians. The scholars of Confucianism Roger Ames and Henry Rosemont Jr. have long held that the role-centered, tradition-bound, communitarian model of the self that we find in Confucianism could serve as an important corrective to the excessive individualism, alienation, and materialism that characterize modern Western societies. While we don't want to fall into the trap of caricaturing ourselves, there is definitely something to this argument, which is why we've seen the rise of broader movements to reenvision the worth of traditional rituals. The *New York Times* columnist David Brooks, to take another example, has argued that our modern celebration of bluntness and straight talking has made us blind to the moral function played by old-fashioned civility and manners. Sounding like a commentator on the Confucian *Analects,* Brooks notes that "smart people who've thought about this usually understand that the habits we put in practice end up shaping the people we are within."

These habits also shape the people around us. The philosopher Hagop Sarkissian has described the Confucian ritual strategy as consisting of a huge number of "minor tweaks in our own behavior—such as our facial expression, posture, tone of voice, and other seemingly minor details of comportment—[that] can lead to major payoffs in our moral lives." This is because of "ethical bootstrapping," the idea that cultivated behaviors have a small positive effect on others, which causes them to act in an incrementally more morally positive way, which in turn feeds back on us. From the perspective of academic psychology, it has become increasingly clear that seemingly trivial aspects of the social and physical environment can have profound effects on behavior. This means that paying attention to the music your kids listen to, what they wear, and whom they hang out with might do both them and society a lot of good. Not a news flash to conservatives, but maybe a bit of a wake-up call to liberals such as myself.

Basic manners also fit into this category of fundamentally important social phenomena we tend to overlook. When I moved to Vancouver from California, one of the first things that struck me was that the natives, when exiting public buses through the back door, always shouted out a loud and cheerful *Thank you!* to the bus driver. It initially struck me as a bit excessive, but I've since come to see it not only as an expression of a fundamentally more pleasant populace—Canadians really *are* nicer than Americans, or at least coastal, big-city Americans—but also as a ritual that probably helps to *create* more pleasant people. The bus driver, whether she realizes it or not, feels better having been thanked; she is now more inclined to drive courteously, or to remain at the stop that extra second to allow someone running late to hop on the bus. This behavior ripples out across my rainy city in subtle waves, much like the mysterious power of *de* described by Confucius, inclining people toward virtue like wind blowing over the grass.

Moreover, work in cognitive psychology suggests that submerging people in a particular cultural tradition also helps them learn to love something they do not already love. Simply exposing someone repeatedly to a new stimulus—a font style, song, painting—causes them, over time, to develop a liking for it. Familiarity breeds love, not contempt. It also makes us more likely to believe in it (for better or worse). Statements that we have heard repeated many times are judged to have a higher "truthiness" quotient than novel statements; the same effect is found with text printed in a familiar or more legible font. This suggests that, for instance, religious rituals that are initially viewed as onerous and strange, or classical texts that at first seem forbidding and difficult, can eventually come to be embraced with spontaneous joy and accepted as valuable and true, simply through an intensive course of training. When this happens in a group setting, the result is precisely the sort of social cohesion that Confucius wanted to encourage.

This has immediate, practical implications for how you go about arranging your daily life. The early Confucians put an enormous amount of effort into modifying their immediate aesthetic environment—clothes, colors, layout of living spaces, music—so that it would reflect the values of the Confucian Way. Although most of us no longer embrace the Way, we can use the same techniques to foster our own particular set of values. If you can set up your home and workplace, to the extent you have control over it, to reflect your tastes and values, the things that make you feel good and at home, you're going to be better off. You'll have more *wu-wei* and more *de*. Colors, landscapes, Led Zeppelin blacklight posters, pictures of your family, religious icons, whatever—environmental reminders of your larger framework of values will reinforce your commitment, and in the process foster absorption, relaxation, and confidence.

You can imagine other situations, too, in which the carving-and-polishing strategy might work best. Dieting, for example, requires transforming a predilection for sausages or scones into a genuine love of leafy greens. And there are other areas where we might want to develop a new disposition: becoming a more patient spouse or parent, a more courteous driver, or a more helpful friend. The basic idea is simple. You choose a desirable model, then reshape your hot cognition to fit by immersing yourself in reminders and environmental cues. How this repetition eventually causes the new internal disposition to become sincere and self-activating is a bit of a mystery—intellectually, the paradox remains—but it seems to work in practice.

There are times, though, when cultural training devolves into empty posturing or when intensive effort turns into counterproductive drudgery. That's when we might need to follow Laozi's "do nothing" strategy.

A growing literature in the psychology of perception has demonstrated that, when it comes to certain difficult visual tasks—exercises where subjects are asked to locate a target shape in the midst of a large array—simply relaxing and letting the answer "pop out" works much better than actively trying. Similarly, when one is stymied by a problem, simply leaving it alone and doing something else is often the best way to solve it. Doing nothing allows your unconscious to take over, and, as we've seen, the unconscious is often better at solving certain types of particularly complex problems. In the field of psychotherapy, so-called "acceptance" strategies, in which clients are instructed to simply allow unwanted thoughts or memories to flood over them, or march before their mind's eye, often prove to be more effective than active suppression strategies. The conscious mind has limited capacity, and often the best thing to do when you run into difficulty is shut it down for a while and let the body take over. Laozi at one point advises the ruler, "Governing a large state is like cooking a small fish"—in other words, you don't want to overdo it. When you are faced with a difficult management decision or an intractable technical problem, the best approach may be just to walk away. Sleep in, take a walk, go weed your garden.

Trying too hard also tends to backfire in social situations. In chapter 4 we talked about the "instrumental" passages in the *Laozi*—the sections that seem practically, and somewhat sinisterly, designed to help rulers and others get ahead in the world. What we see there seems to be advice about how to *fake wu-wei* in order to get its benefits. Whenever I read these *Laozi* passages I'm reminded of a controversial book that came out in the mid-1990s called *The Rules: Time-Tested Secrets for Capturing the Heart of Mr. Right*. The advice could have been lifted directly from the instrumental sections of the *Laozi* or Sunzi's *Art of War*. Retreat in order to advance. Attract interest by pretending not to be interested. In other words, try to create artificial *de*. Part of the controversy that arose around *The Rules* can be attributed to the patently hypocritical and manipulative aspect of the advice; faking interest levels and emotions bugs people, for reasons that should now be obvious. Moreover, despite the enormous amount of money the book has made for the authors, as well as the cottage industry that's sprung up to train people in "the Rules," it's not at all clear that it works very well. Similar guidelines that have been created for men—"the System," for instance, authored by the eminent "Doc Love"—seem equally unsuccessful. The strain of actively trying not to try eventually shows, and it is enormously unattractive when we see it.

The beginnings of agriculture and origins of large-scale human societies are topics unfortunately passed over in silence by Doc Love, but they are arguably much more essential for understanding romantic rejection than anything contained in the System. You can't fake *de*. Human beings are certainly very good liars, but we're also super cheater detectors, and for very good evolutionary reasons that explain everything from the "paradox of virtue" in four-thousand-year-old Chinese oracle bone inscriptions to why large numbers of people in our society remain unhappily single despite having easy access to "the Rules" or "the System." And this is all as Laozi would predict. "The highest Virtue does not try to be virtuous, and so really possesses Virtue"; if you need to talk about benevolence, you're not benevolent.

When it comes to things like dating, or job interviewing, or any situation where the impression you make is important, it's probably best to embrace the uncarved block. If you can follow Laozi's advice and refrain from trying too hard, it's almost inevitably going to go better for you. Fire won't burn you, wild beasts won't attack you, and you may get a second date. The counterproductive nature of trying in such situations was understood intuitively by the great sage and musician Jonathan Richman, who possesses considerably more insight into the human heart than Doc Love. In a classic Modern Lovers track, he addresses obnoxious "bellbottom bummers"— stylish, shallow men who try hard to pick up women but end up getting rudely blown off. He urges these hipsters to remember the example of Pablo Picasso, who exerted an effortless, yet magnetic, pull on women. When he strolled through town or walked into a café, "girls could not resist his stare / Pablo Picasso never got called an asshole."

As a historical account, this is debatable (Picasso was not the nicest guy in the world and probably got called all sorts of colorful things by women), but as a description of the power of *de* it can't be beat. True *de*, true attractiveness, comes out of sincere absorption into a larger, valued good—artistic creation, muffin baking—not calculated clothing choices or pickup lines. What you choose to embrace doesn't matter, as long as it's something that you're doing genuinely, *not* for strategic reasons. There is nothing worse than a wine appreciation class full of single men and women who couldn't care less about what they are tasting: the smell of desperation is so thick it makes it hard to discern the vanilla notes in the Chardonnay.

This has implications for your personal life, to be sure, but it also extends to corporate culture and even national self-identity. It's fashionable these days for corporations to hire "reputation management consultants"—basically, spin doctors and publicists whose job is to create *de* for companies. As one recent article noted, however, there's very little evidence that these services work, because the actual behavior of the companies in question, or their perceived underlying motive, seems to swamp the reputation management efforts. A large petrochemical corporation, for instance, has been running a high-profile ad campaign to soften resistance in British Columbia to a pipeline planned to bring Albertan tar sands oil to the coast. The full-page ads feature comforting pastel colors, frolicking orcas, and quietly competent tanker pilots, but there's a relentlessly cheerful, Potemkin village–like feel to them that rings false. I, personally, would be much more reassured if the company published a detailed, sober, and independently verified risk assessment—or if its past safety record were a little better. Parallel efforts to reassure Americans about the safety of piping Albertan crude across ecologically sensitive Nebraskan sandhills have similarly fallen on deaf ears. When it comes to convincing people that you mean well, talking the talk is never as effective as walking the walk. Evidence suggests that the best way to acquire an enduringly positive corporate reputation is to simply *be* an enduringly positive company.

Nations have an equally hard time manufacturing *de*. A recent report on "the art of soft power" notes that the Chinese government—infuriated that Mickey Mouse and Hollywood movies and other manifestations of American culture continue to overshadow five thousand years of Chinese culture, even in China itself—has embarked on several campaigns to try to raise the coolness quotient of ancient Chinese culture. Two counties in China have recently opened, or are planning to open, enormous, incredibly expensive theme parks organized around Sunzi, the author of the *Art of War*. The idea is that the mystical appeal of this ancient Chinese sage will overpower the attraction of Disneyland Shanghai, set to open nearby in 2015. My money's on Mickey. Ironically, the very culture that gave us the *Laozi* seems to have forgotten the most important of its lessons: if you're *trying* to be popular, you've just ensured your own failure. Actively working to enhance Chinese cultural "soft power"—essentially, political *de*—through top-down government funding or decree is entirely counterproductive. When it comes to soft power in East Asia, the greatest success stories are phenomena like Japanese anime and manga, or the South Korean wave of soap operas and K-pop, that have arisen organically, and unpredictably, from popular culture. It's hard to imagine the ineffably goofy dance moves of "Gangnam Style" being created by a committee of bureaucrats.

THE PARADOX OF INTROSPECTION

The psychologists Jonathan Schooler, Dan Ariely, and George Loewenstein preface a recent paper on happiness with a quotation from Nathaniel Hawthorne: "Happiness is as a butterfly, which, when pursued, is always beyond our grasp, but which, if you will sit down quietly, may alight upon you." This alludes to a phenomenon, supported by previous research, whereby directly pursuing certain outcomes, like happiness or pleasure, can have a counterproductive effect. The authors present two new studies suggesting that both the active monitoring of pleasure and the deliberate intention to enjoy an activity actually lead to decreased enjoyment. At first glance this may seem to support the Laozian strategy: sit on your uncarved block and wait for the butterfly of happiness to arrive.

The authors, however, conclude with a more complicated message. They identify what they call the "paradox of introspection," a tension that seems to exist between effortful reflection and the attainment of pleasure and happiness. While it's clear that both active pursuit and excessive reflection (on, say, the taste of wine or jam) can have a disruptive, negative effect on sensory pleasure, the complete absence of introspection can leave you unable to recognize the elements of the experience that truly bring you joy. Revealingly, the negative effects of introspection disappear once one reaches a certain level of expertise: intellectual analysis and a rich descriptive vocabulary eventually become assets for both jam experts and wine professionals.

This is where the Mencian strategy of mixing effort and naturalness seems particularly helpful. Just as food or wine connoisseurship draws upon innate tastes but refines and extends them, morality and happiness may be amenable to gentle cultivation. We've all experienced moments of insight during conversations with spouses, close friends, or therapists. Like King Xuan in the ox story, we often have only imperfect access to our own motivations. Part of Mencius's job is to help the king, through introspection, to identify the real motive behind his sparing of the ox, learn to accurately identify this feeling again in the future, and then strengthen and extend it.

This whole process works much better, and proceeds faster, when guided by an experienced coach or teacher. The best way to refine your wine appreciation abilities is to sample a large variety of wines in a reflective manner, checking your intuitions against those of an expert and then reevaluating your experience in light of their comments. The result is a refining feedback loop, where the expert advice helps to focus and redirect your perception (*I never thought to look for pencil lead, but sure enough there it is*), giving you a broader and more subtle descriptive vocabulary, which in turn serves to open up previously unnoticed aspects of your experience.

When it comes to moral development, this could take the form of the kind of dialogue that we see in the *Mencius*, with King Xuan being talked through his experiences and receiving specific tips on how to recognize his sprout of compassion. Or, for those who lack the services of a personal moral trainer, heading to the library or picking up a Kindle might be the solution. The psychologist Jonathan Haidt observes that no less a figure than Thomas Jefferson argued for the moral function of great literature, defending the inclusion of novels—then viewed as vulgar and potentially morally dangerous—in his own personal collection at Monticello: "When any . . . act of charity or of gratitude, for instance, is presented to our sight of imagination, we are deeply impressed with its beauty and feel a strong desire in ourselves of doing charitable and grateful acts also. On the contrary, when we see or read of any atrocious deed, we are disgusted with its deformity, and conceive an abhorrence of vice. Now every emotion of this kind is an exercise of our virtuous dispositions, and dispositions, like limbs of the body, acquire strength by exercise." Just as the four limbs of the body are strengthened by exercise, so are the four sprouts of proper *wu-wei* behavior nourished by the imaginative workout provided by literature. One wonders if Jefferson had been reading Mencius.

We can of course obtain the same effect from theater or film, which—like Confucian music—seems to reach directly inside us to alter our emotions. For instance, there is some evidence that watching empathy-inspiring video clips increases generosity and caring behavior. All of this should give pause to school reformers inclined to water down or eliminate literature and arts programs in favor of more "practical" training in math, science, or economics. Unless we are happy producing nothing more than hyper-rationalist Mohist sprout-pullers, the arts are crucial for engendering socially desirable forms of *wu-wei*.

In addition to one-on-one therapy and immersion in edifying art forms, various types of guided meditation might help develop sprout-grounded *wu-wei*. The philosopher David Wong cites a neuroimaging study of a group of sixteen expert meditators (defined as those who had at least ten thousand hours of meditation practice under their belts) who were asked to engage in loving-kindness-compassion meditation, which involves focusing on the image of a particular loved one, allowing feelings of compassion to "invade the mind," and then trying to extend this feeling to "all beings." After meditating, subjects were exposed to audio recordings with a range of emotional valences: positive (baby laughing), negative (woman in distress), and neutral (background restaurant conversations). Compared to novice meditators, the experts showed a markedly amplified response to the negative sound in parts of the brain (particularly the insula cortex) associated with emotion processing and empathy. This suggests that compassion might very well be something like a sprout that can be developed, or a muscle that can be strengthened, through imaginative training or meditation.

In an attempt to apply these insights to people who may not have ten thousand hours to spare, researchers at Emory University have piloted a program of "cognitive-based compassion training" with elementary school and foster care children in the metropolitan Atlanta area. This training, which takes place twice a week over the course of eight to ten weeks, is a secularized form of traditional Tibetan Buddhist meditation practices, and it looks like something that could have benefited King Xuan. Preliminary observations suggest that it has a significant effect on children's ability to both experience and express compassion.

Mencius does not emphasize the role of Confucian ritual in his sprout-development therapy, but it is possible that structured, formalized practices other than meditation could support the growth of our moral sprouts. Happiness researchers have found that individuals asked to perform random acts of kindness one day a week reported overall increases in reported happiness and that simple acts of generosity, such as giving away $5, can increase subjective feelings of happiness. Similarly, although forcing oneself to perform a compassionate act does not immediately make one compassionate, it's likely that, say, volunteering in a soup kitchen regularly might over time strengthen one's "sprout of empathy," especially if the work is accompanied by imaginative reflection. Related work by the psychologist Robert Emmons and colleagues suggests that keeping a gratitude journal—which forces one to reflect upon the positive aspects of one's life—improves physical and mental health and leads to increased compassion for others.

Nonetheless, there are certainly other times when analysis and introspection are decidedly counterproductive. We've seen that focusing conscious awareness on the mechanics of one's performance, while useful in very early stages of skill acquisition, has a disruptive effect on more experienced players or performers. Similarly, regardless of level of expertise, focusing on the environment and effects one wishes to have upon it ("external focus") is more effective than focusing on one's own bodily movements or internal states ("internal focus"). For instance, swimmers told to focus on pushing the water back (external focus) as opposed to pulling their hands backwards (internal focus) swim faster, and this effect has been shown in a large variety of domains. There are various hypotheses about why directing one's attention out, rather than in, is more effective in learning and performing a physical skill, but it seems likely that it has to do with *wu-wei.* When you focus on your own movements, you allow your conscious mind to insert itself where it doesn't belong, disrupting smooth, automatic motor programs and allowing other distractions—social pressure, personal anxieties, promised material rewards—to invade and degrade your performance. Focusing on the skill-relevant environment facilitates your ability to get "lost" in the to-and-fro of the play.

It's in these situations—when the pressure is on and the stakes are high—that Zhuangzi's letting-go strategy really shines. Woodcarver Qing has to engage in a weeklong ritual before he can successfully forget his body and focus his skill on the environment, and elsewhere there are suggestions that seated meditation and breathing practices are involved. It is likely that these Zhuangzian exemplars are engaged in something that looks more like the "objectless" meditation practiced in Zen, Zhuangzi's direct religious heirs, than the visualization practices of Tibetan Buddhism. Objectless meditation typically involves assuming a simple, stable posture and then trying to empty your mind through focusing on your breath or a fixed feature in the environment. A growing empirical literature on this type of meditation suggests that it has a positive effect on mental concentration, reaction time, motor skills, and perceptual sensitivity to the environment. This is probably because meditation downregulates the conscious, cold-cognition centers of our brain, thereby creating room for hot cognition to do its thing.

Importantly, meditation has also been shown to increase self-esteem, empathy, trust, and other traits crucial to interpersonal relations, and of course this is really the goal of Zhuangzian fasting of the mind: allowing us to be in *wu-wei*, and possess *de*, when interacting with others. In social situations, having an "external focus" would mean turning your attention to the personalities, conversations, and body language around you, rather than focusing on yourself. Actually *caring* about the conversation instead of reflecting on whether you can contribute to it or consciously monitoring how people are reacting to you is what's really important. The key is to go with the flow and be moved by the environment rather than trying to control it. Meditation might help you enter this state; a vigorous bout of exercise or a shot of vodka might do the same. What works best for you probably has to be worked out through trial and error.

TAKING THE BODY SERIOUSLY

I've argued in this book that the phenomena of *wu-wei* and *de* are central to human flourishing and cooperation. The only reason we need to be told this is that recent Western thought has been so obsessed with disembodied rationality that embodied spontaneity—along with the unique tensions it presents—has fallen off the radar. Thinking of moral perfection as a matter of following rules or calculating utility certainly simplifies things. Reason carefully, throw in a bit of willpower, and you're done. The problem is that this model is deeply wrong. It's psychologically unworkable, given what we know about the way the human body-mind operates. Moreover, it completely fails to reflect how we actually experience our lives.

Things are slowly changing. Scientists have, in recent decades, begun moving away from abstract models of human cognition toward more embodied ones. They're coming to recognize that the sort of knowledge that we rely on most heavily is hot, emotionally grounded "knowing how" rather than cold, dispassionate "knowing

that." We're made for *doing*, not thinking. This has significant implications for everything from how we educate people to how we conduct public debates, make public policy decisions, and think about
our personal relationships. There is a lot to be said for trying to take
a disembodied, abstract stance to the extent that it's possible for
humans: many trends can be picked out only by applying statistical
tools to a huge amount of data. When it comes to essentially political decisions, however, the idea that different options can be boiled
down into mathematical equations is not only misleading but posi-
tively harmful. We think in emotion-laden, embodied images, and
as long as we remain oblivious to these "metaphors we live by" we
remain imprisoned by them.

Since the early Chinese philosophers aimed for an action-
oriented model of perfection, they focused on training the embodied mind through physical practice, visualization exercises, music,
ritual, and meditation. There was little emphasis on abstract theo-
rizing or the learning of general principles. Although memorization
played a role—students were expected to know the classics by heart
at an early age—the end goal was learning to use this information
in real life, flexibly and creatively. Confucius once noted, "Imagine a
person who can recite the several hundred Odes by heart but, when
delegated a governmental task, is unable to carry it out or, when sent
abroad as an envoy, is unable to engage in repartee. No matter how
many Odes he might have memorized, what good are they to him?"
Simply memorizing the classics does not make one a true gentleman or lady—you need to *incorporate* this knowledge, make it part
of your embodied being. This is what early Chinese training focused
on. The goal was to produce a kind of flexible *know-how*, exemplified
in effective engagement with the world. Education should be analog,
holistic, and oriented toward action.

In terms of our personal lives, Mihaly Csikszentmihalyi and his
colleagues have performed an invaluable service by focusing attention on the role of "flow" states in human happiness. It needs to be
recognized, though, that the qualities that have been attributed to

flow depend less on complexity and challenge than on absorption in a valued whole. No doubt there are those who, because of their personalities, can best attain *wu-wei* by climbing a more challenging rock face or closing a more complex deal. *Wu-wei* is also found, though, in tending your garden, walking down an easy path in a landscape that you love, or spending unstructured playtime with your child. Moreover, it's only this kind of value-grounded spontaneity that gives rise to *de*, an attraction that arises from subtle bodily signals. Acknowledging the role of *de* in human trust and cooperation not only helps us to understand why the paradox of *wu-wei* exists in the first place but also gives us new insight into the dynamics of human relationships. It explains why trying to be attractive makes you unattractive, and why trying to be cool just makes you look silly.

It's worth wondering how the intellectual trajectories of Asia and Europe would have differed if the Mohists, or some other early Chinese rationalist school, had won out over the Confucians, or if the more embodied Aristotelian model had remained dominant in the West. The abstract, rationalist strategy certainly has its advantages. For one thing, it was central to the development of Western science, which in turn powered the technological and economic revolutions that resulted in European ships appearing in China to establish colonies instead of the other way around. A frightening amount of academic ink has been spilled on the question of why modern science arose in the West and not in China, which for most of its history was much more technologically advanced, more populated, and better educated. There is probably no single decisive reason, but it's plausible that an important factor was a deep-seated suspicion of abstract thought for its own sake and a corresponding failure to develop a disembodied, instrumental stance toward the world.

Despite a lot of griping among intellectuals about the dangers of modern scientific alienation, I count myself as a fan. I find the idea of living in a world without antibiotics, electric lights, airplanes, running hot water, and surgical interventions incredibly unappealing. It's ironic, though, that disembodied scientific rationality has, in recent years, begun to reveal the profound limits of abstract thought when it comes to our lived experience. We have very good scientific reasons for thinking that cold logic—especially when we're talking about one human being relating to another, or a person learning to inhabit his or her social world—is extremely useful in an indirect way but can't actually be trusted behind the wheel. This is why the more embodied, integrated model of perfection we find in early China is worth recovering and dusting off.

Our modern conception of human excellence is too often impoverished, cold, and bloodless. Success does not always come from thinking more rigorously or striving harder. In a world increasingly dominated by cram schools, treadmills (literal or otherwise), 24/7 connectivity, and punishing amounts of stress, seeing the world in terms of the power and grace of spontaneity can help us to make better sense of our work, our goals, and our relationships. The paradox of *wu-wei*, unfortunately, is going nowhere—there is inevitably a tension involved in trying not to try. In practice, though, there is often a way through. The awkward first moments of a blind date can morph into an evening of spontaneous conversation and real connection. Pregame jitters dissolve as we become absorbed in the excitement and challenge of the game. Even salsa dancing can be fun. Our lives are full of these moments of unselfconscious, effortless enjoyment. If we can manage to not push too hard when trying is bad, and not think too much when reflection is the enemy, the flow of life is always there, eager to pull us along in its wake.

Acknowledgments

In many ways this book represents a distillation of almost twenty years of thinking about early China and the intersection of Chinese thought and modern science, so fully documenting my influences and intellectual debts is impossible. I apologize in advance for any omissions, of which there are no doubt many.

My dissertation on the paradox of *wu-wei* (Slingerland 1998) was inspired by David Nivison's work on the "paradox of virtue" and was pursued under the guidance of my advisers at Stanford, P. J. Ivanhoe and Lee Yearley. Since in this book I've tried to keep scholarly notes to a minimum, I refer the reader to my academic treatments of these topics (Slingerland 1998, 2000, 2003b, 2008) for complete references to the scholarly literature that I am building on. Thanks to Cynthia Read and Oxford University Press (New York City) for permission to reuse my translations from Slingerland (2003b), and Deborah Wilkes and Hackett Publishing Company for permission to reuse translations from Slingerland (2003a).

In thinking about connecting the Chinese conception of *wu-wei* to modern cognitive science, I was very fortunate that the eminent French neuroscientist Alain Berthoz chose to spend part of 2009 as a Distinguished Visiting Professor at the Peter Wall Institute for Advanced Studies at the University of British Columbia (UBC). Alain took an immediate interest in *wu-wei* and its relevance to his own work on embodied action, and our informal chats led to two workshops on the *wu-wei*–cognitive science connection hosted by the Collège de France and the Peter Wall Institute. I am very much indebted to Alain, both institutions (especially the former Peter Wall director Dianne Newell), and the workshop participants (Jon Elster, Romain Graziani, Anne Cheng, Jean-Luc Petit, Brian Bruya, Jim Enns, Ron Rensink, Todd Handy, Pierre Zakarauskas) for the work we did on thinking through the significance of *wu-wei* in a contemporary context. Jim was particularly helpful in introducing me to his research on the "automatic pilot." The extraordinarily fertile intellectual community at UBC also brought me into contact with a visiting cognitive psychologist, Rolf Reber, whose interest in the paradox of *wu-wei* in the *Analects* led me to his very relevant and important work, as well as to a coauthored piece on the topic (Reber and Slingerland 2011). Rolf also gave me some helpful feedback on this manuscript. Finally, my thoughts on the Chinese philosophy–modern science connection have benefited very much from academic panels and conversations over the past five or six years with colleagues such as David Wong, Owen Flanagan, Brian Bruya, Hagop Sarkissian, and Bongrae Seok. David's and Hagop's work in particular has given me new insights into the psychological sophistication of the Confucian strategies, and Matt Bedke has provided important general philosophical advice.

My turn toward the cognitive sciences would have been much less productive without the guidance of my colleagues in the UBC psychology department, who have without exception been incredibly generous in sharing their time and expertise. In particular I would like to thank Jonathan Schooler (since departed for the University of California Santa Barbara), Liz Dunn, Toni Schmader, Joe Henrich, and Ara Norenzayan. Effusive thanks to Kalina Christoff for watching my neuroscience back. Others in the field, particularly Jon Haidt, Dan Wegner, and Sian Beilock, have also been extremely generous with both advice and references. I owe a particular debt to Jon for showing me (with *The Happiness Hypothesis*) that one could write a popular book without having to sacrifice intellectual content and while still making a contribution to the professional literature. And I cannot emphasize enough that any remaining errors in terms of the empirical literature remain my own; friends can do only so much. Thanks too to Eric Margolis, Mark Collard, and Harvey Whitehouse for advice on the general project.

Having been persuaded by my UBC colleague and friend Ara Norenzayan that writing a popular book was a great idea, and possessed of an extremely vague idea for one, I had to pass through the scrutiny of my eventual agent, Katinka Matson at Brockman, Inc. Katinka was critical about what needed to be critiqued, was ruthless in transforming my meandering academic-ese into something readable, and guided me with sure hands through the strange world of trade publishing. This project would never have happened without her help.

Much of this book was written while on a sabbatical in Rome, partly supported by a UBC Killam Research Fellowship. Thanks to the Killam Trust, as well as to the Canada Research Chairs Program, from which I derived teaching release and research support after returning from leave. Thanks too to the folks at Enoteca Il Piccolo, where a fair amount of this manuscript was written or polished, as well as Moves palestra in via dei Cimatori, which kept me sane, allowed me to work off all the wine and pasta, and was the place where Emma Lo Bianco gave me insight into the paradox of *wu-wei* from an actor's perspective and pointed me toward some relevant literature. My old friend Andrea Askowitz—a pretty good tennis player and wonderful prose stylist—gave me some helpful feedback on the introduction.

I am very grateful to those friends, relatives, and colleagues who heroically made it through the entire manuscript. Joseph Bulbulia—a scholar, gentleman, and great friend—saved me from some serious conceptual and scientific blunders, and my father-in-law, Donald Lammers (an eminent retired academic in his own right), gave me some pointed and useful comments. My mother-in-law, Giovanna Colonelli Lammers, a Latin, French, and Italian teacher, helped me to untangle some convoluted Italian syntax. Extraordinary thanks are owed to Todd Keithley, an old friend, successful New York City attorney, and former literary agent extraordinaire, who somehow found time in his busy life to give me detailed feedback on the entire manuscript, covering everything from line edits to global stylistic tics to deep conceptual problems. If you occasionally run across a bit of unusually lucid and direct prose, it's probably because Todd was involved in either editing whatever convoluted version I'd come up with or simply suggesting a better version in the margins. Any remaining stylistic crimes are purely the result of my own laziness or stubbornness.

It would be impossible to overstate the role played in creating what you have just read by my editor at Crown, Amanda Cook. "This isn't feeling very *wu-wei* to me," was Amanda's disappointed comment about the first draft of this book, and she was unfortunately quite right. Four chapters that I produced in a frenzy of writing during the beginning of my sabbatical—and that I was quite happy with—were ruthlessly boiled down to one, which under her guidance was then further combined with another subsequent chapter, and another after that. Amanda took the sloppy soup of my first attempt to write a popular book and refined it into something that we're both fairly confident people will enjoy reading. Her excellent sense of pacing and style, combined with a profound knowledge of the scientific literature, great suggestions for contemporary examples, and an unerring sense of what to cut and what to keep, helped me to produce a book infinitely superior to what I would have concocted on my own. If this were a scientific publication, Amanda would be listed as a coauthor, but the conventions of mainstream book publishing limit me to thanking her profusely for everything she has done to encourage and shape this project. Thanks also to Amanda's assistants, Emma Berry and Domenica Alioto, for all of their help; in particular, Emma's final edits of the manuscript were extremely perceptive and useful.

Finally, and as ever, my thanks to my wife, Stefania Burk, and daughter Sofia, who dealt with an often harried and distracted husband/father during the time I spent working on this book. Writing about effortlessness is harder than you would think, it turns out. They have been my primary source of blissful spontaneity for the last decade. And to my brother-in-law, Greg Burk, with whom I've had endless discussions about spontaneity and jazz, fueled by Roman sun, great food, and Italian wine, I say, *wu-wei* out.

Appendix

Much of this book is dedicated to describing the strategies for attaining *wu-wei* and *de* developed by a group of early Chinese thinkers; the table below provides a basic scorecard of who is who to help you keep track of the discussion.

EARLY CHINESE THINKERS AND TEXTS

THINKER(S)	SCHOOL	RELEVANT TEXTS	STRATEGY
Confucius/Xunzi (*shuoon-tzuh*)	Confucianism	The *Analects*, the *Xunzi*	"Carving and polishing"; try really hard for a really long time
Laozi (*lao-tzuh*)	Daoism	The *Laozi* or *Daodejing*	"Uncarved block"; stop trying immediately, go home
Mencius	Confucianism	The *Mencius*	"Cultivate the sprouts"; try, but don't force it
Zhuangzi (*juang-tzuh*)	Daoism	The *Zhuangzi*	"Let go"; try to forget all about trying or not trying, just go with the flow

As explained in notes to the Introduction, the terms *Confucianism* and *Daoism* are a bit anachronistic but helpful in that they pick out two very broad strategies for cultivating *wu-wei*: trying (education, cultural training) and not trying (de-cultivation, forgetting).

Notes

INTRODUCTION

1   *"Don't play the saxophone. . . ."* Quoted in Safire and Safir (1989: 79).
2   *The actor Michael Caine.* Caine (1990: 28–29).
2   *A 2005 piece in* Sports Illustrated. *Sports Illustrated* (2005).
3   *Precisely* for *having lost their stuff.* One of the most vivid, excruciating accounts of this struggle that I've ever read is actually a fictional one, from the novel *The Art of Fielding* (Harbach 2011). The protagonist, Henry Skrimshander, is a budding superstar college shortstop, aggressively courted by agents and scouts, who suddenly loses his mojo, and much of the novel centers on the paradoxical tensions involved in getting oneself back into the zone. At the height of his anxiety, for instance, Henry sits alone in the middle of the night at his customary spot between second and third base, his head spinning with all of the advice that seems so useless: relax, stop thinking, be yourself, be the ball, stop trying so hard. "You could only try so hard not to try too hard before you were right back around to trying too hard," Henry muses. "And trying hard . . . was wrong, all wrong" (305).
3   *Blass was eventually forced into early retirement.* Hattenstone (2012). Blass has recently published a memoir about his experiences (Blass and Sherman 2013). A related example from a few decades earlier is the famous meltdown of Ralph Guldahl. Guldahl was a young golfing superstar in the late 1930s, exploding on the scene with U.S. Open victories in 1937 and 1938 and then winning the Masters in 1939. After writing a book about the mechanics of golfing, however, he never won another major championship. Some speculate that Guldahl just started *thinking* too much about golfing to actually golf well (Collins 2009; thanks to Eric

Margolis for bringing this example to my attention). In attempting to explain the nature of Guldahl's mental block, Collins cites the inimitable Yogi Berra's comment, "How can you hit and think at the same time?" He ultimately concludes, however, that more mundane concerns may be responsible for Guldahl's decline. In any case, it's clear that conscious reflection has a negative effect on expert performance, a theme that we'll explore in some detail below. Also see recent helpful books on the phenomena of performance and choking—both on the field and off—from a former world-class athlete (Syed 2010) and a psychologist (Beilock 2010).

3   *She later managed to make a professional comeback.* Simon quotation from Holden (1987). Simon's comeback was apparently enabled by the development of distraction strategies to get her past her nervousness, at least one of which involved getting spanked on the bottom by band members immediately before performing (Simon spanks away her stage fright 2006). Whatever works.

7   *Confucian and Daoist schools.* See the Appendix for a table listing the thinkers we'll be looking at, their school (Confucian or Daoist), and the texts that convey their teachings. Caution needs to be observed when talking about school affiliation, especially during the Warring States period. You will often hear that "Daoism" is an ancient Chinese school of thought, founded by Laozi and continued by his follower Zhuangzi. This is anachronistic and philosophically inaccurate. *Daoism* is a term invented by later scholars to organize the Han imperial library and was not used at all by the people who authored the *Laozi* or the *Zhuangzi*. There was no self-identified "Daoist" school in the Warring States, and—as we will see—the thought of Zhuangzi is different in many ways from what we see in the book attributed to Laozi. Nonetheless, I'll continue to use *Daoist* to describe Laozi and Zhuangzi because it's a widely known term and because the two thinkers do share important features (the scholars who put them together under the same rubric were no dummies). It should be noted that my more finicky colleagues have recently also begun avoiding the term *Confucian* to refer to early thinkers such as Confucius, Mencius, or Xunzi. The term used by Warring States thinkers like Xunzi to refer to themselves, which is often rendered as "Confucian" (*ru*, 儒), originally meant simply "erudite" or "scholar," and Confucius himself did not emerge as a great cultural hero until well after the Warring States period. I'm less worried about talking about "Confucians" than I am about "Daoists," however,

since both Mencius and Xunzi self-identified as followers of Confucius and his Way. If you're a stickler for historical accuracy, feel free to mentally replace "Confucian" with "Ru-ist" as you read this book.

7   *Integration of the body, the emotions, and the mind.* The individual may still at times pause to weigh various options or consider the situation ahead, but even this reflection is performed with a kind of effortlessness. So *wu-wei* is not "mindless" behavior but rather behavior that springs directly from embodied thinking.

8   *No other language has a good equivalent to either* wu-wei *or* de. Because both of these words are difficult and awkward to translate, I'm going to simply refer to them in their phonetically spelled-out form.

9   *"Warring States" period.* Technically, the beginning of the Warring States period is marked by the death of Confucius, but it is certain that the text that contains his teachings, the *Analects*, was not composed until after his death in the early Warring States period. I'll also be using names like "Laozi" and "Zhuangzi" to refer to the author(s) of the received texts that bear these names, whether or not it's historically likely that these figures actually existed or composed the texts in question. See the Appendix for an overview of the thinkers and texts, as well as a pronunciation guide.

10   *Why spontaneity and effectiveness hang together.* On "fast and frugal" heuristics, see Gigerenzer (2002); on the power of the unconscious mind, see Wegner and Bargh (1998); Wilson (2002); Kahneman (2011).

10   *Snazzy fMRI diagrams.* The images produced by fMRI analysis are schematic representations of a highly mediated product of complex, and sometimes controversial, statistical techniques. For a discussion of the limits to our ability to infer actual cognitive function from fMRI, see Poldrack (2006).

11   *Increasingly confirmed by work in cognitive science.* This literature will be reviewed in chapter 6.

11   *Certain contemporary ideas of spontaneity miss important aspects of the* wu-wei *experience.* As we'll see, modern science is catching up and has stumbled onto a set of problems that look remarkably like the early Chinese paradox of spontaneity: If you are not already spontaneous and unselfconscious, how can you *try* to get there? As cognitive scientists have become more interested in spontaneity and nonconscious modes of behavior, they have begun to perceive this tension as well and (as scientists are wont to do) to attack it with the tools of their

trade. Social psychologists interested in the effectiveness of unconscious processes—the power of the "blink" moment, to borrow the title of Malcolm Gladwell's popularization of this research (Gladwell 2005)—have also stumbled upon the deleterious effects of overthinking. Prompted by the Russian proverb that, once so instructed, it is difficult *not* to think of a white bear, psychologists have been exploring phenomena that look very much like the paradox of *wu-wei*: how difficult it is to consciously suppress a thought, force oneself to relax, or will oneself to sleep (Wegner 2011). There is a growing experimental literature focusing on the phenomenon of "choking" under pressure, which also provides some intriguing hints about what factors in particular contribute to the phenomenon so dreaded by professional athletes and performers (Beilock 2010). Cognitive neuroscientists have also begun to probe the human brain to uncover what's going on under the hood, as it were, when athletes experience the "runner's high" (Dietrich 2003) or when jazz musicians switch from scale playing to free improvisation (Limb and Braun 2008). This work provides us with a helpful contemporary lens through which to view the ancient paradox of *wu-wei*.

12   *It tends to portray rational thought as the essence of human nature.* There are some exceptions worth noting: for instance, Aristotle in ancient Greece, or more recent thinkers such as Friedrich Nietzsche, the phenomenologist Maurice Merleau-Ponty, and American pragmatists such as William James or John Dewey.

12   *Some sort of mind-body dualism seems to be a human psychological universal.* See Bloom (2004) for a good survey of the empirical evidence for "folk" mind-body dualism and Slingerland (2013) for an argument that even the early Chinese embraced a weak or "sloppy" form of mind-body dualism, although one much less extreme than what we see in most post-Enlightenment Western thought.

12   *This led them down some very unproductive paths.* For instance, early cognitive scientific work treated thought as completely "amodal" (that is, having nothing to do with concrete images or bodily perceptions) and as occurring in the "mind," a ghostly substance connected in some mysterious way to the body (Descartes thought the connection ran through the pineal gland, but he didn't find a lot of takers for this theory). Intellectually steeped as they were in the disembodied model, it took cognitive scientists a while to figure out how deeply wrong it was. However, their great advantage over philosophers—who still, for the most part, haven't heard the news—is that they do experiments and gather empir-

ical data. That is, they don't just accept models, they actually test them against the world. As philosophers of science are only too happy to tell you, it is notoriously difficult to get a straight answer from such test-ing, but we can know *some* things, and the inability of the disembodied model to account for how human beings actually think and act happens to be one of them. Realizing that they'd been stuck with a bum philo-sophical model, in the 1980s and 1990s scientists started looking around for alternatives.

13   *Human thought as fundamentally "embodied."* For a helpful introduc-tion to the issue of so-called "first-generation" versus second-generation cognitive science and the embodied cognition movement, see Gibbs (2006); on philosophical efforts to "put the body back in the mind," see Johnson (1987).

13   *We inevitably think in terms of journeys and paths not taken.* On the per-vasiveness of metaphor in our lives, see Lakoff and Johnson (1980, 1999).

13   *The embodied view of cognition.* On the imagistic nature of thought, see Kosslyn, Thompson, and Ganis (2006); on emotions and reason, see Damasio (1994) and Berthoz (2006); on the action-oriented nature of thought, see Gibson (1979) and Noë (2004).

13   *Melding of cognitive science and Chinese thought.* See, for example, Varela, Thompson, and Rosch (1991); Thompson (2007); Flanagan (2011).

13   *Important corrective to the tendency of modern Western philosophy.* For some examples inspired directly by work in cognitive science or evo-lutionary psychology, see Munro (2005); Bruya (2010); Slingerland (2011a); Seok (2012). For related (but less empirically grounded) work inspired by American pragmatism and embodied movements in recent Western thought, see the work of Roger Ames and Henry Rosemont Jr. (e.g., Rosemont and Ames 2009).

13   *Know-how.* For classic work on "knowing that" versus "knowing how," see Ryle (1949) and Polanyi (1967).

14   *What real human excellence looks like.* A helpful way to think about *wu-wei* is as the embodied alternative to the modern Western, disembod-ied ideal of perfection. Exaggerating only slightly, the ideal person for most recent forms of Western thought would be a disconnected brain floating in a vat, free of all bodily-induced emotions and distractions. This creepy floating brain would be able to draw in information from the environment through electronic sensors, process it, calculate the best rational decision, and then send out orders to some equally creepy robotic devices that would carry out its will. This all seems so distaste-

ful because it's not *natural*; we feel at some level that it's not really how we work. This is where popular culture gives us insight into something that our expert traditions have tended to miss: in any sort of novel or movie, the dispassionate, rational, instrumental manipulator—the ideal Western philosophical agent—is always the *villain*. We distrust this ideal, and find it vaguely repulsive when it's laid out in exaggerated form, because we realize at some level that we are not disembodied brains and that our limbs are not mere mechanical devices. We are integrated, well-tuned, body-mind systems, capable of relatively dispassionate thought but guided primarily by intelligent emotions, perceptual habits, and spontaneous impulses. This is why the hero, who always in the end defeats the evil villain, looks a lot like someone in *wu-wei*: brave but not reckless; passionate but not foolishly so; clever but not manipulative.

15   *"A collection of clerics, misogynists, and puritan bachelors."* Baier (1994: 114). This collection of essays includes several influential pieces exploring the neglect of trust and other forms of implicit, emotional cooperation in recent Western thought.

17   *Rule-bound, impersonal public institutions.* For introductions to this debate in light of some recently discovered archaeological texts, see Cook (2004) and Slingerland (2008).

1. SKILLFUL BUTCHERS AND GRACEFUL GENTLEMEN

19   *"At every touch of his hand . . ."* *Zhuangzi*, ch. 3 (B. Watson 1968: 50). All of the translations from the Chinese will be my own and in many cases are drawn from my own previous works (Slingerland 1998, 2003a, 2003b). To help the reader track down the context of the passages, I'll also give a chapter number and page reference to a recommended English translation.

20   *"When I first began cutting up oxen . . ."* B. Watson (1968: 51).

20   *"A skilled butcher has to change his cleaver once a year. . . ."* B. Watson (1968: 51).

21   *"Whenever I come to a knot . . ."* B. Watson (1968: 51).

22   *"When I am getting ready . . ."* *Zhuangzi*, ch. 19; B. Watson (1968: 205–6).

22   *"Now I set off for the mountain forest . . ."* *Zhuangzi*, ch. 19; B. Watson (1968: 205–6).

22   *All he then had to do was cut away the stone that didn't belong.* An ex-

cerpt from a sonnet attributed to Michelangelo himself reads: *"Non ha l'ottimo artista alcun concetto / Ch'un marmo solo in se non circo-scriva / Col suo soverchio, e solo a quello arriva / La man che obbedisce all'intelletto"*: roughly, "No artist, no matter how excellent, possesses an idea that is not already contained in the marble block, imprisoned by superfluous matter that only the hand guided by the intellect can remove" (Buonarroti 1821: 1; thanks to Giovanna Lammers for help with the translation).

23   *"The Way of Heaven / Excels in overcoming, though it does not contend..."* Laozi, ch. 73; Ivanhoe (2003: 76).

23   *"Do not go out the door, and so understand the whole world..."* Laozi, ch. 47; Ivanhoe (2003: 50).

24   *What is less widely appreciated, however, is that ... effortless ease and un-selfconsciousness ... also plays a central role in early Confucianism.* I make this case in Slingerland (1998, 2003b).

24   *"The Master said, 'At fifteen I set my mind upon learning...'"* Analects 2.4; Slingerland (2003a: 9).

25   *In the early Chinese accounts of wu-wei described above ...* Actually, in most of these early Chinese stories, the term *wu-wei* does not appear at all. In these stories, the hallmarks of *wu-wei* are conveyed using a wide variety of other terms or metaphors: *relaxing, following, forgetting, flow-ing.* The reason that *wu-wei* eventually emerged as a general blanket term for these sorts of stories and experiences is that it is the most ab-stract expression of a general sense of letting go or unselfconscious ease (Slingerland 2003b).

26   *This split-self talk is clearly metaphorical rather than literal.* The meta-phor theorists George Lakoff and Mark Johnson have documented this basic tendency of languages to distinguish between a "Subject" (the "I," the locus of identity and consciousness) and one or more recalci-trant "Selves"—either a vague entity ("myself") or a specific body part or bodily function ("my tongue," "my emotions"). The *Subject-Self* re-lationship is metaphorical in the sense that there is always really only *one* person involved, me, despite the fact that I might experience myself metaphorically as if I were two (Lakoff and Johnson 1999: 268–70). It is this Subject-Self dichotomy that helps us to understand why the phrase *wu-wei*—"no doing" or "no trying"—could be used as a general label for what otherwise seems like very active engagement with the world. *Wu-wei* refers metaphorically to a state where some action is oc-curring in the world but the "Subject"—the conscious "I"—is not exert-

ing force or actively focusing attention on it. Whatever is happening is happening without the Subject's active involvement.

26   *A typically colorful example from the* Zhuangzi. Zhuangzi, ch. 17; B. Watson (1968: 183–84).

27   *There is now general agreement that human thought is characterized by two distinct systems. . . .* An important early articulation of a two-system or dual-processing model of cognition in modern psychology is found in Sloman (1996), although the terms "System 1" and "System 2" were originally coined by Stanovich and West (2000). For the general reader, Daniel Kahneman, perhaps the best-known figure in dual-processing research, has written a wonderfully thorough and accessible introduction to the topic (Kahneman 2011). Most recently, Ap Dijksterhuis and colleagues have proposed distinguishing as "System 3" or "Type 3" the cognitive processes involved in making important, complicated life decisions, which we typically make after mulling things over or letting them sit for long periods of time (Dijksterhuis et al. 2013). These processes have a consciously determined goal (like System 2) but work largely unconsciously (like System 1) and tend to be very slow (like System 2, only slower). Dijksterhuis proposed summarizing these three types of cognition as "blink" (System 1), "think" (System 2), and "sleep on it" (System 3). Further research is needed to determine if System 3 constitutes a set of processes worth distinguishing as a separate category or merely represents one particular profile of interactions between System 1 and System 2.

29   *They don't want to shake his hand.* Thumbtack experiment originally performed by Edouard Claparede, as reported in LeDoux (1996: 181–82); for general discussions of explicit versus implicit cognition, see LeDoux (1996); Zajonc (1980); Pessoa (2005).

29   *Unconscious "knowing how" seems distinct from conscious "knowing that."* For classic philosophical discussion of "knowing how" versus "knowing that," see Ryle (1949); for representative modern scientific discussions, see LeDoux (1996), work on "two-stream" visual processing (Goodale and Milner 2004), or research on what my colleague at the University of British Columbia James Enns calls the body's "automatic pilot" (Enns and Liu 2009).

30   *A chess master sees lines of force, zones of danger, and paths of opportunity.* Chase and Simon (1973).

31   *Japanese Zen Buddhism is derived from Chinese Chan Buddhism.* Zen is just the Japanese pronunciation of a Chinese character meaning "meditation," which is pronounced *chan* in modern Mandarin.

32   *A direct line of intellectual descent between the story of Butcher Ding in the* Zhuangzi *and that of Luke Skywalker in* Star Wars. The influence of Zen on Lucas appears to have been primarily through the ideal of *bushido,* which in turn was mediated by the films of Kurosawa (Baxter 1999: ch. 7).

32   *"For three months after, he did not even notice the taste of meat. . . ."* Analects 7.14; Slingerland (2003a: 68).

33   *The* oomph *that is the hallmark of conscious will or effort.* Wegner (2002: ch. 1).

33   *"Stroop task."* This variant of the Stroop task is adapted from Kahneman (2011: 25).

33   *The "cognitive control regions" of the brain.* Although much of the work on cognitive control has focused on the lateral PFC (LPFC), which seems to play a key role in the cognitive regulation of behavior, it's clear that other parts of the PFC (such as the ventromedial PFC [VMPFC]) are also important in emotional and bodily regulation. For some helpful survey articles on cognitive control, see Miller and Cohen (2001); Banich (2009); Stout (2010); Braver (2012).

34   *Squabbling sets of neurons.* Interestingly, although the lateral PFC seems to be the region actually exerting control, the ACC appears to be the source of the feeling of conscious *oomph* that accompanies mental effort. Lionel Naccache and colleagues report the case of a patient with a large left hemisphere frontal lesion, including the ACC, who was able to perform tasks, like the Stroop task, that require cognitive control but who reported no feeling of subjective effort: he could *exert* cognitive control but lacked the *oomph* that normally goes with it. Other studies have also suggested that the ACC is very much linked to subjective feelings of effort and difficulty (Naccache et al. 2005).

35   *A clever study by Charles Limb and Allen Braun.* Limb and Braun (2008).

35   *The medial prefrontal cortex (MPFC).* Not a lot is known about the function of the particular portion of the medial PFC that was activated in this study, but the area in general seems to be crucial in storing and retrieving information about the self. Some early fMRI work on the medial PFC found that it was differentially activated when subjects were asked to think about themselves ("Does this character trait describe you?") as opposed to others ("Does this trait describe George Bush?") (Kelley et al. 2002). Other work on subjects practicing "mindfulness" meditation reveals a decrease in activity in the lateral PFC but increased activation in the ACC and medial PFC, suggesting a sort of

flexible openness, grounded in the unconscious self, that circumvents the narrow spotlight of the conscious mind (Cahn and Polich 2006; Siegel 2007). Limb and Braun, the authors of the jazz study, speculate that decreased activity in the lateral PFC, combined with continued ACC and medial PFC activation, may result in a kind of "defocused, free-floating attention that permits spontaneous unplanned associations, and sudden insights or realizations" because the cognitive control centers of the mind are no longer actively regulating the contents of one's consciousness (Limb and Braun 2008: 4).

36   *The kind of relaxed but vigilant mode we enter into when we're fully absorbed in a complex activity.* Some more recent work on cognitive control has referred to this state of readiness as "reactive" cognitive control, as opposed to "proactive," where we are experiencing a high degree of subjective effort and both the lateral PFC and ACC are very active (Braver 2012).

36   *The body unleashed is impressive to behold.* It's important to realize that the body that takes over in activities like jazz improv is not just any body but one that has been previously shaped and trained by the conscious mind. This tension between training and relaxation is at the heart of the controversies we'll be discussing below.

36   *Evolution has off-loaded the vast bulk of our everyday decision making . . . onto our automatic, hot, unconscious system.* The psychologist John Bargh has given the very provocative estimate that 99.44 percent of what we do is the result of unconscious, hot thought (Bargh 1997: 243); this is comically precise, but the relevant percentage is certainly quite high.

36   *An enormous, and growing, popular literature on the power of unconscious thought.* Wilson (2002); Gladwell (2005); Kahneman (2011); Duhigg (2012).

36   *"Thinking without thinking."* Gladwell (2005). Also see Ap Dijksterhuis et al.'s work on "deliberation without attention" (Dijksterhuis et al. 2006), as well as a helpful recent review by Bargh and colleagues of automaticity in social cognition (Bargh et al. 2012).

37   *"The choice of notes, the silences, the attacks . . ."* CD liner notes from *The Way In*, 483 Music, 2006 (Copyright 482 Music and Greg Burk). In addition to being an inspired musician, Greg is my brother-in-law and has found the concept of *wu-wei* the perfect expression of his own artistic process. Be sure to get the CD, which includes the track "Wu-wei Out."

2. DRUNK ON HEAVEN

38 *"When a drunken person falls out of a cart ..."* Zhuangzi, ch. 19; B. Watson (1968: 198–99).

39 *"His spirit is intact."* Zhuangzi, ch. 19; B. Watson (1968: 198–99).

40 *"One who rules through the power of* de *is like the Pole Star."* Zhuangzi 2.1; Slingerland (2003a: 8).

41 *The Laozian sage is invisible....* See Ivanhoe (1999) on the differences between Confucian and Laozian political *de*.

42 *Wu-wei, on the other hand, means becoming part of something larger: the cosmic order represented by the Way.* See Graham (1983: 9–13) on the objective nature of spontaneity in China, as opposed to more subjective Western conceptions.

42 *The example of mastering a physical skill ... becomes misleading if it's disengaged from its original cultural and religious context.* The argument that *wu-wei* is a fundamentally religious ideal is made at length in my 1998 PhD dissertation, more easily accessible through the article (Slingerland 2000a) and book (Slingerland 2003b) versions.

43 *The concept of "flow."* This concept entered popular consciousness with the publication of *Flow: The Psychology of Optimal Experience* (Csikszentmihalyi 1990), but Csikszentmihalyi had actually been exploring flow with teams of collaborators since the 1960s, establishing himself as one of the seminal figures in what has come to be known as the "positive psychology" movement. While other approaches in academic psychology tended to focus either on mental illness or on the narrow mechanics of cognitive functioning, positive psychology sought to distinguish itself by its more holistic approach and ambition to use psychological research to enhance human flourishing. Csikszentmihalyi's work, in particular, forced psychologists to start paying attention to the power of what we've been calling the "embodied mind": not abstract rational thought or blind physical reflexes, but rather action arising from the effortless and unselfconscious harmony of both body and mind. Csikszentmihalyi himself is aware of parallels between flow and early Chinese accounts of what I am calling *wu-wei*, and he even recounts the story of Butcher Ding in his 1990 work. An interesting side-note is that his son, Mark, was my classmate at Stanford, where we studied early Chinese thought together and where I was first exposed to the idea of flow. Mark is now a distinguished professor of Chinese thought at the University of California Berkeley.

43   *"Spiraling complexity" that "forces people to stretch themselves...."* Csik-
      szentmihalyi (1988a: 30).

43   *"A scholar of international reputation..."* Csikszentmihalyi (1990: 32).

44   *The story of Serafina Vinon...* Csikszentmihalyi (1990: 146).

46   *Most flow experiences actually occur in social situations of relatively low
      complexity.* See, for instance, the essays collected in Csikszentmihalyi
      and Csikszentmihalyi (1988b).

46   *These researchers are, for the most part, Western individualists.* Interest-
      ingly, studies of flow by some of Csikszentmihalyi's non-Western col-
      leagues have tended to focus more on social experience and shared
      values: see, for instance, Ikuya Sato's study of *bosozuku* motorcycle
      gangs in Japan (Sato 1988).

47   *We go beyond the facts anytime we make value judgments.* The distinction
      between "is" and "ought" is a venerable topic of Western philosophical
      speculation and is far beyond the scope of this book. For an introduc-
      tion, see David Hume's original essay (Hume 1739/1888: bk. 3, pt. 1, sec. 1),
      G. E. Moore's work on the related "open question" (Moore 1903), and dis-
      cussion of the two in Sayre-McCord (2012). This dependence of human
      beings on value commitments that go beyond the facts is an important
      element of human psychology missed by the more rabid among the "New
      Atheists," who believe that humans can be guided solely by empirical, sci-
      entific evidence and utilitarian calculation. (An added irony is that these
      champions of science appear completely ignorant of basic work in the
      cognitive science of religion or moral psychology.) I'd single out the work
      of Sam Harris as the most philosophically simplistic and empirically un-
      informed example of this genre (e.g., Harris 2004, 2010).

47   *We cannot do without such convictions.* For the "inescapable" nature of
      moral frameworks, see Taylor (1989, esp. chs. 1–3); for more recent work
      on the nature of "secular" society, see Taylor (2007). For a less technical
      introduction to Taylor's views and how they look from a scientific per-
      spective, see Slingerland (2008: ch. 6) or a video of my talk "Confucius
      as Secular Savior" (Slingerland [2007]).

47   *The distinguishing feature of* wu-wei ... *absorption of the self into some-
      thing greater.* See Slingerland (2000a, 2003b) for more on this distinc-
      tion between flow and *wu-wei* as a fundamentally religious ideal; in the
      last few years some younger scholars have also noted the parallels and
      distinctions between flow and *wu-wei* (see, e.g., De Prycker 2011; Bar-
      rett 2011), although not always with reference to previous work on the
      topic.

48   *"Sacred," or endowed with supercharged meaning.* For an account of reli-
      gion as a "making special," grounded in more classic work in religious
      studies but also informed by contemporary cognitive science, see Taves
      (2009).

48   *Secular humanism functions very much like a traditional religion. It gives
      its followers a set of sacred values.* American conservatives therefore have
      a more accurate view of the world than their liberal compatriots in at
      least one regard: values like freedom and human dignity *are* part of a
      broad value system that is incompatible with other sets of values run-
      ning around in the world. American liberals, on the other hand, tend to
      think that embracing liberal values is simply a human default position
      that all people are inexorably drawn into once they are freed of false
      consciousness. The claim that modern secular humanism, despite its
      own conceit of having transcended frameworks entirely, still functions
      very much like a traditional religion in supporting value judgments has
      long been made by the eminent Canadian philosopher Charles Taylor
      (1989, 2007). Taylor's claim is backed empirically by his own histori-
      cal work, the research of sociologists such as Robert Bellah and col-
      leagues (Bellah et al. 1996), and the current state of the art in moral
      and social psychology (see Haidt 2012 for a recent review). For an edify-
      ing and somewhat amusing spectacle of leading philosophers and psy-
      chologists (metaphorically) gouging each other's eyes out debating this
      issue, see the video proceedings of a 2007 conference called "Beyond
      Belief," held at the Salk Institute in Southern California (http://the
      sciencenetwork.org/programs/beyond-belief-enlightenment-2-0). This
      will also give you the chance to see me getting beat up by a gang of New
      Atheists.

49   *Umbrella of secular humanism.* One of the historically unprecedented
      features of modern Western life is the fragmentation of meaning that
      results from the embrace of multiple, relatively narrow personal frame-
      works, which typically leads to a kind of incoherence when individ-
      uals try to articulate or justify their values. In a well-known analysis
      of modern American life, *Habits of the Heart,* the Berkeley sociologist
      Robert Bellah and colleagues documented the fact that many modern
      Americans have trouble coherently defending their moral intuitions
      because they draw their specific day-to-day guidelines from a variety of
      sources and often even lack a clear sense of the overarching framework
      of secular humanism that encompasses them (Bellah et al. 1996).

54   *The collected teachings of Confucius.* The *Analects* is clearly a composite
        text that began to be assembled only after the death of Confucius him-
        self, although there is scholarly debate about how precisely to date its
        various chronological layers (see Slingerland 2000b for a discussion).
        The vast majority of the text seems to be quite early, perhaps roughly
        contemporaneous with our next text, the *Laozi*, and we'll be treating
        it as a reasonably unified book reflecting the thought of the very early
        Confucian school.

54   *How Confucius behaved when receiving a blind master musician as a guest.*
        Analects 15.42; Slingerland (2003a: 190).

56   *A young man named Yuan Rang. Analects* 14.43; Slingerland (2003a: 172).

56   *His follower Xunzi.* Mencius of course also saw himself as a follower
        of Confucius, but—as we'll see in chapter 5—his approach to culti-
        vating *wu-wei* differs in some significant ways. Xunzi's approach looks
        much more like something of which the historical Confucius would
        approve, and Xunzi himself (writing at the very end of the Warring
        States period) criticized Mencius as a "fake" Confucian. The fact that
        Confucius and Mencius, rather than Confucius and Xunzi, are typi-
        cally paired together today is the result of later historical events, having
        little to do with the actual landscape of Warring States Confucian
        thought. That said, although I'll be conflating Confucius and Xunzi
        in this chapter for the sake of convenience, they did have some quite
        different views on a variety of topics, including the nature of Heaven
        (*tian*) and the origin of the Confucian Way. For instance, Confucius
        saw the Way as something that had been revealed by Heaven to the
        Zhou kings. Xunzi, on the other hand, saw *tian* not as a god but as a
        neutral force (perhaps better translated as "nature"). He portrays the
        Way as a kind of cultural artifact, cobbled together over a long period
        of time by a series of moral geniuses. (See Ivanhoe 2007a on differ-
        ent views of Heaven in Confucius and Xunzi.) In this respect, Xunzi's
        account of the origins of large-scale societies—how we developed the
        cultural technologies to move us from the state of nature to civiliza-
        tion—is a bit more empirically plausible, especially if you see this accu-
        mulated cultural innovation as a sometimes unguided process, so I'll be
        primarily working with Xunzi's conception of the Way in this chapter.

57   *"I once engaged in thought for an entire day without eating. . . ." Analects*
        15.31; Slingerland (2003a: 172).

58 *"I once stood on my tiptoes . . ."* Xunzi, ch. 1 (Knoblock 1988: 136).

59 *Recent scientific work offers some answers to this question.* For the past several decades, the field of psychology has tended to focus on the power of the unconscious. The intellectual pendulum, however, has begun to swing the other way. Recently, some important voices have reemphasized the importance of the conscious mind and have pushed back against suggestions that consciousness is a mere "epiphenomenon"—a by-product of the *real* mind, the unconscious mind, as it goes about its business. A recent review article by the psychologist Roy Baumeister and colleagues, entitled "Do Conscious Thoughts Cause Behavior?," perhaps best sums up current thinking on the functions of consciousness within an embodied perspective (their answer to the question posed in their title is a cautious *yes*); see Baumeister, Masicampo, and Vohs (2010).

59 *Complex modeling of other minds.* Dennett (1991).

59 *Offline, imaginary practice is equally important when it comes to physical skills.* See, for example, Grouios (1992).

60 *Without the sense of narrative unity that it provides we'd be completely at sea.* For an accessible neuroscientific account of the role of narrative in the construction of self-identity, see Gazzaniga (1998).

60 *"Scaffolding."* Dennett (1995: 379–80).

60 *This scaffolding includes such useful cultural inventions as, say, calculus. . . .* On how cold, statistical reasoning outperforms hot cognition, or *wu-wei* dispositions, in a variety of situations, see especially Kahneman (2011: 239–41).

60 *The creation of modern science is essentially a story of how, over a long period of time, humans have cobbled together novel methods of thinking. . . .* McCauley (2011).

61 *A body of information passed down from generation to generation.* Culture understood in this way includes a wide variety of information in one's social environment, including physical instruction and objects (modeled skills, measuring tools, books, a culturally transformed environment) and verbal communication (on culture understood in a naturalistic manner, see, e.g., Sperber 1996). A lot of work has been done recently on the process of cultural evolution, whereby human cultures in all of their various incarnations develop over time in a process that shares characteristics with genetic evolution but also differs in important ways (on cultural evolution, see esp. Richerson and Boyd 2005). Culture is transformative: these evolved cultural systems shape us in

profound and typically unnoticed ways, altering our implicit associations, our basic motivations, our preferences and tastes, and even our
spatial cognition. On the foundational role of culture in human cognition, see Geertz (1973) for a more traditional anthropological account
and Henrich, Heine, and Norenzayan (2010) for an updated, evolutionarily informed one.

62   *The result is an adaptive set of food taboos.* On Fijian taboos, see Henrich
and Henrich (2010); also see Henrich and McElreath (2003) and Katz,
Hediger, and Valleroy (1974) for more examples. Thanks to Joe Henrich for detailed feedback on this topic.

62   *The "collective mind" of any cultural group . . . is typically smarter than any
individual human mind.* Boyd, Richerson, and Henrich (2011).

62   *Xunzi compares the Confucian Way . . . to markers used to indicate a ford.*
Xunzi, ch. 17 (Knoblock 1994: 21).

62   *You should probably just shut up and do it.* It's significant that, compared
to chimpanzees, human children are extremely accurate mimics: if you
show them a new skill, they will reproduce every step of what they've
seen, even the nonessential ones, like the fact that the model scratched
her nose in the middle of the procedure. Chimps will ignore a lot of the
details and just imitate the causally effective steps, which is more efficient when causal efficaciousness is easily discernible (Nagell, Olguin,
and Tomasello 1993). It's not as useful when it comes to the kind of complex cultural information that human beings transmit to one another,
where the very complexity makes it extremely difficult, if not impossible, to separate the essential from the nonessential steps. "Monkey-see,
monkey-do" actually describes *humans*, not any other ape, and that's
probably a result of the specific evolutionary selection pressure that
learning human culture exerts upon us.

64   *"Pay attention!"* The pun is from Schmeichel and Baumeister (2010), a
helpful and recent review of the attentional control cost literature.

64   *"Ego depletion."* Baumeister et al. (1998); Muraven, Tice, and Baumeister
(1998). More recent work exploring the probable physiological underpinnings of this phenomenon suggests that the availability of glucose
to the brain seems to be the rate-limiting factor: exerting cognitive control depletes glucose, and giving subjects supplemental glucose can help
them recover self-control more quickly (Gailliot et al. 2007). As Gailliot et al. note, the glucose-greedy nature of cognitive control has interesting implications when it comes to certain acts of will, in that dieting
simultaneously involves exertion of willpower and restriction of glucose

intake. The result may be an "ironic conflict" in which the dietary restriction produces lower glucose, which, in turn, undermines the willpower needed to refrain from eating.

65   *The answer lies in* domestication. This focus on domestication is true more generally of so-called "virtue ethical" approaches to moral education and decision making, as opposed to the cognitive control models that have dominated recent Western thought. See Haidt (2005); Slingerland (2011a).

65   *Consider what happens when we learn a new physical skill.* Example taken from Dietrich (2004: 752–53); see this piece for an excellent, and quite readable, description of some of the neurological mechanisms underlying *wu-wei* or flow states.

66   *"Basal ganglia memory."* The basal ganglia function something like a gateway or triggering mechanism at the center of a wide range of networks that implement behavior; see Gazzaniga, Mangun, and Ivry (1998: 78–81, 300–307).

66   *The online control of our actions is shifted to them.* See Berthoz (2006: 103–6) for more on the neuroscience of automatization.

66   *The basal ganglia and relevant motor regions can essentially learn the skill on their own.* Grafton, Hazeltine, and Ivry (1995). It's also likely that degeneration of the basal ganglia is responsible for the inability of patients with Parkinson's and Huntingdon's diseases to acquire and retain motor skills, even when their conscious memories are relatively unaffected (Grahn, Parkinson, and Owen 2009).

66   *Brain activity gradually drops.* M. Smith, McEvoy, and Gevins (1999); Poldrack et al. (2005).

66   *The conscious mind can gradually relax.* The beauty of this arrangement is that our limited cognitive resources—previously swamped by the demands of steering and braking and accelerating—are now freed up for other tasks. Crucially, this allows for a degree of flexibility and innovation not available to the nonexpert. Having more or less mastered the basics of driving an automatic transmission, I could not only devote more attention to considering which Led Zeppelin blacklight poster to buy at the mall but also merge more effectively and smoothly into traffic and cope with unexpected conditions, such as a patch of ice or some aggressive Jersey driver cutting me off. There are, to be sure, limitations to downloading skills into hot cognition: automated motor routines take time to retrain when conditions change. After driving an Italian rental car with manual transmission for several months, it usually takes me a

week or two to stop stomping on the floor with my left foot—feeling for a nonexistent clutch pedal—when I'm home and back behind the wheel of my automatic. I do eventually stop, though, and the cost of that week or two of minor adjustment is more than outweighed by the benefits of automatization. There's simply no way I could have navigated the unique perils of Roman driving if I'd also had to be constantly, consciously thinking about how to operate a clutch.

67   *How human beings moved from a state of nature to civilized life.* Xunzi describes our original state, before civilization, in terms very much like what the Western philosopher Thomas Hobbes (1588–1679) called "the war of all against all" that existed before we created governments and states (Hobbes 1651/1985: 189). In the state of nature, every individual is out to satisfy his or her own needs, with no concern for any kind of public good, the result being universal suffering. Both Hobbes and Xunzi viewed the state of nature from what is probably an excessively individualistic perspective: we are, in fact, pack animals, so the original war was really the war of family against family, or clan against clan.

67   *"The ancient sage kings viewed this chaos with revulsion. . . ."* Xunzi, ch. 19 (Knoblock 1994: 55).

68   *Each of these roles comes with a defined repertoire of proper behaviors.* For more on the function of social roles in Confucian ethics, see Ames (2011); to be inoculated against some of Ames's more extreme views concerning the radical difference of Chinese conceptions of the self, see the review of this book by Ihara and Nichols (2013).

69   *"In summer, he wore a single layer of linen or hemp. . . ."* Analects 10.6, 10.12, 10.26; Slingerland (2003a: 100–101, 105, 109).

69   *A sweaty jogging outfit.* In fact, the line cited above, "In summer, he wore a single layer of linen or hemp but always put on an outer garment before going out," essentially means that Confucius wouldn't go out in public wearing only a T-shirt, an etiquette rule not at all uncommon in, say, contemporary Europe or Latin America.

70   *Small gestures, tone of voice, and facial expression can change the mood of your social surroundings.* The classic treatment of this topic is Goffman (1959); also see Sarkissian (2010) and Wegner (2002: esp. 193–94) on unconscious, interpersonal cueing.

70   *Babies are born with an innate ability to smile. . . .* Stern (1977), discussed in de Sousa (1987: 182–83).

70   *This process is often encouraged by stories, art.* Another way to put it is that this maturation of emotional response occurs under the guidance

of what the neuroscientist Ronald de Sousa refers to as "paradigmatic scenarios." See de Sousa (1987).

71   *Someone "standing with his face to a wall."* Analects 17.10; Slingerland (2003a: 204–5).

72   *A particular historical moment in Silicon Valley.* See Jobs's biography for details of the influence of his social and intellectual environment on his later success (Isaacson 2011); also see Yglesias (2011) for a brief discussion of Jobs's biography that emphasizes this point. As Yglesias observes, "It's just obvious from reading the early chapters of the book that there could have been no Apple Computer if not for the fact that Jobs was born and raised in Silicon Valley. Not just in the sense that Jobs was personally fortunate to have been in the right place at the right time, but in the sense that the place itself had to exist for the right combination of building blocks to be put into place."

72   *The bumpers in bowling-alley gutters.* For more on the importance of situational buffers in early Confucianism, see Hutton (2006) and Slingerland (2011b).

72   *"Do not look unless it is in accordance with ritual. . . ."* Analects 12.1; Slingerland (2003a: 125).

73   *The lyrics of the Zheng music—preserved in the* Book of Odes. Refer to the *Book of Odes,* "Airs of Zheng" (nos. 75–95).

73   *Subjects who unscrambled a word jumble . . .* Bargh, Chen and Burrows (1996).

73   *Subjects who unscrambled sentences about helpfulness . . .* Macrae and Johnston (1998). For reviews of the relevant literature, see Bargh and Chartrand (1999) and Wilson (2002). Quite recently controversy has arisen because of repeated failed attempts to replicate previously published results (e.g., Pashler, Coburn, and Harris 2012), and there are now calls to rigorously review the entire priming literature (Yong 2012). Some of the failed replication attempts suggest that priming does, in fact, work, although not quite in the way suggested by earlier research. For instance, a very interesting recent study (Doyen et al. 2012) failed to replicate the precise results of Bargh, Chen, and Burrows (1996), except in a condition where the "old person" primes were delivered in the presence of an experimenter who *expected* the subject to walk more slowly as a result; this suggests that conceptual priming has a causal effect on behavior when accompanied by subtle interpersonal clues. This combination of conceptual priming and interpersonal cueing would, of course, be characteristic of Confucian education.

73   *Subjects primed by the social role "professor" performed significantly better . . . , while subjects primed with the "soccer hooligan" role performed more poorly.* Dijksterhuis and Van Knippenberg (1998). For a review of some of this literature, see Dijksterhuis and Bargh (2001).

73   *Subjects holding a pen in their teeth . . .* Strack, Martin, and Stepper (1988).

74   *Ritual behavior does indeed have an immediate feedback effect. . . .* Alcorta and Sosis (2007).

74   *"Its ability to enter inside and pluck at the heartstrings."* This comes from one of the so-called "Guodian texts" to be discussed in chapter 7 (*Xing Zi Ming Chu*, strip 22).

75   *People are quite accurate at identifying certain emotions . . . motivating a piece of music.* Indeed, children as young as three or four seem able to do this at better-than-chance accuracy, and the trick seems to be universal acoustic patterns—rhythm, key—that correspond to basic human emotions. For work on this and the effect of music on emotions, see the literature review in Juslin and Västfjäll (2008).

75   *The social cohesion created by music and dance is the very reason they are so common.* Durkheim (1915/1965). In a passage that sounds as if it could have come from an early Confucian text, the anthropologist Roy Rappaport similarly observes that, when dancing and singing, the members of a religious community experience the truths taught by their faith "not only through their ears and eyes, but coming out of their own bodies in song, or forcing entry into their bodies through the beat of drums animating their limbs in dance" (Rappaport 1999: 388).

76   *Synchronized group activities like music, dance, and marching help communicate emotions. . . .* McNeill (1995); Wiltermuth and Heath (2009); Konvalinka et al. (2011).

76   *The ineradicable tragedy at the heart of modern life.* Freud (1930/1969).

77   *"The learning of the gentleman enters his ear . . ."* *Xunzi*, ch. 1; Knoblock (1988: 140).

77   *"What need has he, then, for strength of will . . ."* *Xunzi*, ch. 21; Knoblock (1994: 108). Xunzi even compares perfected ethical behavior to such paradigmatic hot processes as perception or basic motor programs: "He cultivates the model of the ancient kings as easily as telling white from black; he responds perfectly to situational changes as effortlessly as counting from 'one' or 'two'; he puts into action the basic rules of ritual in a perfectly easy manner, as if he were merely moving his four limbs" (*Xunzi*, ch. 8; Knoblock 1990: 76).

79   *"I will not open the door for a mind that is not already striving...."* Ana-
      lects 7.8; Slingerland (2003a: 66).

79   *"I have yet to meet a man who loves* de ..." *Analects* 9.18; Slingerland
      (2003a: 92–93).

79   *"Is goodness really so far away? ..."* Analects 7.30; Slingerland (2003a: 74).

79   *"Those for whom it is genuinely a problem ..."* Analects 6.12; Slingerland
      (2003a: 56–57).

80   *"Those who try to censure the village poseur ..."* Mencius 7.B.37; Van
      Norden (2008: 195).

80   *"Nowadays 'filial' means simply being able to provide one's parents with
      nourishment...."* Analects 2.7; Slingerland (2003a: 10).

4. STOP TRYING

83   *One of these stories.* Analects 18.6; Slingerland (2003a: 216–17).

85   *Another, similarly strange, passage in the* Analects. *Analects* 14.39; Sling-
      erland (2003a: 170).

87   *Whoever compiled the standard received version of the text.* The tradi-
      tional story of the text's origins holds that the Old Master, disgusted
      with the corruption and excesses of his society, decided to leave China
      for greener pastures but was stopped on his way out by a border guard,
      who refused to let him leave until he recorded his wisdom for poster-
      ity. The roughly five thousand characters that he jotted down became
      the *Laozi.* Some versions of the story claim that he went as far as India,
      where he began teaching and became known under the alternate name
      Shakyamuni—in other words, he became the historical Buddha. This
      version gained popularity after Buddhism entered China in the early
      centuries of the Common Era, and it was used to explain the perceived
      similarities between Buddhism and Daoism. They were similar be-
      cause the Indians got it all from Laozi. This is complete fantasy. Recent
      archaeological discoveries have confirmed the long-held suspicion that
      "the" *Laozi* was in fact a fluid text in the Warring States, with the in-
      dividual passages combined in a variety of ways, probably to suit the
      individual needs of the particular compiler. This is more or less what
      you'd expect from a countercultural, somewhat anarchistic movement:
      no fixed text, free borrowing, lots of diversity. Nonetheless, we'll be
      basing our discussion primarily on the standard received text.

87   *"The court is corrupt...."* Laozi, ch. 53; Ivanhoe (2003: 56).

88   *"The five colors blind our eyes. . . ." Laozi*, ch. 12; Ivanhoe (2003: 12).

89   *"When everyone in the world knows that the beautiful is 'beautiful,' . . ."*
        Laozi, ch. 2; Ivanhoe (2003): 2. This line could also be rendered "When
        everyone in the world knows that the beautiful is 'beautiful,' it is be-
        cause it is ugly" or "When everyone in the world knows that the beauti-
        ful is 'beautiful,' this is when ugliness arises."

89   *"He who speaks does not know." Laozi*, ch. 56; Ivanhoe (2003: 59).

89   *"Verbal overshadowing."* The term *verbal overshadowing* was coined by
        Schooler and Engstler-Schooler (1990).

89   *Tim Wilson and Jonathan Schooler had subjects taste five different types of*
        jam. Wilson and Schooler (1991).

89   *A similar study involved the evaluation of dorm-room posters.* Wilson et
        al. (1993). It is important to note that the negative effects of verbaliza-
        tion seem to disappear with experts (e.g., Wilson, Kraft, and Dunn
        1989; Melcher and Schooler 1996), which might form part of Confu-
        cius's counterattack.

90   *"There is no crime greater than indulging your desires. . . ." Laozi*, ch. 46;
        Ivanhoe (2003: 49).

90   *"The head monkey at Paris puts on a traveller's cap. . . ."* Thoreau
        (1854/1949: 23). He then adds the classic line "I sometimes despair of
        getting anything quite simple and honest done in this world by the help
        of men." Thoreau would have loved the *Laozi*.

91   *"Hedonic treadmill."* For a review, see Frederick and Loewenstein (1999).
        More recent work has qualified some of the earlier, stronger conclusions
        in the field; for instance, it's clear that baseline happiness can indeed
        change over time, that people constitutionally differ in their baseline
        levels of happiness, and that "happiness" itself is a complex phenome-
        non with many different subcomponents, all of which could be moving
        in different directions at any given moment. (See Diener, Lucas, and
        Scollon 2006 for a review.)

92   *How we stack up against our neighbors or colleagues.* The so-called "Eas-
        terlin paradox" is that, although relative income predicts happiness
        within a society at any given time, overall societal increases in income
        have not resulted in increases in happiness (Easterlin 1974). On the
        importance of income rank versus absolute income, see, for example,
        Boyce, Brown, and Moore (2010).

93   *"Grasping the vessel and filling it to the rim . . ." Laozi*, ch. 9; Ivanhoe
        (2003: 9).

95   *"The highest Virtue [de] does not try to be virtuous . . ." Laozi*, ch. 38; Ivan-

hoe (2003: 41); this chapter actually begins the text in one archaeological version, the Mawangdui silk manuscript.

96   *"It is only when the great Way falls into disuse . . ."* Laozi, ch. 18; Ivanhoe (2003: 18).

96   *"Many of our favorite goals, when pursued consciously, can be undermined . . ."* Wegner (2002: 311).

97   *Trying to stop thinking of sex is the best way to think of sex.* On ironic effects, see the literature reviews in Wegner (2002, 2009).

97   *"The Putt and the Pendulum."* See the discussion in Wegner (2009), based upon the original study reported in Wegner, Ansfield, and Piloff (1998).

97   *"Sleep [is like] a dove which has landed near one's hand. . . ."* Frankl (1965: 253), quoted in Ansfield, Wegner, and Bowser (1996).

97   *The result is that they actually fall asleep faster.* Ascher and Turner (1980).

97   *Botching easy shots in the next set.* Reported in Dietrich (2004: 756).

98   *Subjects who are explicitly instructed to be fair and unprejudiced . . .* Monin and Miller (2001).

98   *People asked to imagine that they had just donated time to community service . . .* Khan and Dhar (2007).

98   *The pernicious effect of* labeling *morally positive behavior.* Sachdeva, Iliev, and Medin (2009). Interestingly, those whose moral identities were threatened by negative primes like *greedy, mean,* or *selfish* upped their donations to an impressive average of \$5.30 (reviewed in Sachdeva, Iliev, and Medin 2009).

99   *"Get rid of wisdom, abandon distinctions . . ."* Laozi, ch. 19; Ivanhoe (2003: 19); I am, however, using the older Guodian version of the passage.

99   *"One who engages in learning adds to himself day by day . . ."* Laozi, ch. 48; Ivanhoe (2003: 51).

100   *"Mystery upon mystery . . ."* Laozi, ch. 1; Ivanhoe (2003: 1).

101   *"Carrying on your back your encumbered earth soul . . ."* Laozi, ch. 10; Ivanhoe (2003: 10).

101   *"Transient hypofrontality."* See the literature review in Dietrich (2003).

102   *"Some of the phenomenologically unique features of this state . . ."* Dietrich (2003).

103   *"Demanding nothing in return for his kindness . . ."* Laozi, ch. 81; Ivanhoe (2003: 84).

103   *"The crooked will be whole; . . ."* Laozi, ch. 22; Ivanhoe (2003: 22).

103   *The naturalness of the sage causes others around him to become natural*

as well. See, for instance, *Laozi*, ch. 57; Ivanhoe (2003: 60): "I am *wu-wei*, and the people are naturally transformed / I am fond of stillness, and the people are naturally rectified / I am without action, and the people naturally prosper / I desire not to desire, and the people naturally become like the uncarved block."

104   *"It is reasonable to suppose that in a primitive culture . . ."* Csikszentmihalyi and Csikszentmihalyi (1988a: 184).

105   *"Noble Savage" portrayals of small-scale or exotic societies.* Csikszentmihalyi and Csikszentmihalyi (1988a: 184), citing Redfield (1953). Interestingly, the authors also attribute this kind of constant flow to other ancient cultures, such as ancient Rome, India, or China, where at least the elites might have "moved through life with the intricate grace of ballet dancers, and derived perhaps the same enjoyment from the challenging harmony of their actions as they would have from an extended dance" (186).

105   *Extreme social stratification.* At least as soon as they could accumulate enough wealth to make stratification worth someone's while. There's evidence that hunter-gatherer societies are relatively egalitarian, with built-in social mechanisms for preventing any particular individual from becoming too powerful (Boehm 1999). Once wealth levels get high enough, however—in regions like the Pacific Northwest with great natural resources, or as a result of the invention of agriculture—social stratification seems to immediately follow.

105   *Levels of hunger, disease, cruelty, and violence that we would find shocking.* On the Noble Savage myth in contemporary anthropology, see Horton (1993: 88–97, 133–36); on the idea of small-scale societies as in harmony with nature, see N. Smith (2001); on the myth of small-scale societies as peaceful, see Keeley (1996), Le Blanc (2004), and Pinker (2011).

105   *"Reduce the size of the state and decrease its population. . . ."* *Laozi*, ch. 80; Ivanhoe (2003: 83).

107   *"I declare that those who wish to take the world and do something to it will not be successful. . . ."* *Laozi*, ch. 29; Ivanhoe (2003: 29).

107   *"In the same way, if you desire to rule over people . . ."* *Laozi*, ch. 66; Ivanhoe (2003: 69).

108   *"Most people are unimpressed by the fact that flow provides an optimal subjective experience. . . ."* Csikszentmihalyi (1988b: 374). *Autotelic* refers to an activity that is engaged in with no outside goals other than the activity itself.

109   *"Most people are loud and boisterous. . . ."* *Laozi*, ch. 20; Ivanhoe (2003: 20).

5. TRY, BUT NOT TOO HARD

III   *"In the state of Song there was a man . . ."* Mencius 2.A.2; Van Norden (2008: 40).

III   *"There are those who think that there is nothing they can do . . ."* Mencius 2.A.2; Van Norden (2008: 40).

112   *Lumping him together with Confucius and Xunzi.* Xunzi was the dominant follower of Confucius in the early imperial period, but the thought of Mencius was revived in the twelfth century c.e., when Mencius was declared the orthodox follower of Confucius by the so-called "neo-Confucians." For most contemporary Chinese people, it is therefore Confucius and Mencius that are associated with early Confucianism, with Xunzi remaining relatively unknown. This, incidentally, is why Mengzi/"Mencius" has a Latinized name, like Kongfuzi/Confucius, whereas Xunzi does not: the Jesuits who brought back to Europe our first knowledge of Confucianism in the seventeenth century weren't told about Xunzi.

113   *The maximization of the material wealth, population, and order of the state.* In our contemporary, overpopulated world, increasing the population doesn't sound like a desirable end, but in Mozi's time Chinese territories were characterized by large areas of arable but unused land, so more people meant more agricultural output, as well as a stronger army.

113   *No more lifetime cultivation of refined moral sensibilities.* For a helpful discussion of consequentialism-utilitarianism (Mozi) versus virtue ethics (the Confucians), with particular reference to the Mohist versus Confucian debate, see Van Norden (2007).

114   *It is immoral not to give away all of one's income . . .* Singer (2011: 175–217, 191–95).

116   *"Now, imagine a person turning around and, all of a sudden, spotting a small child . . ."* Mencius 2.A.6; Van Norden (2008: 46).

116   *"Imagine that one is so hungry that a single plate of rice . . ."* Mencius 6.A.10; Van Norden (2008: 154).

117   *Quite a bit of support for some of his basic claims.* As we'll discuss later, this support also has the effect of undermining rationalist models of ethics from ancient Mohism to modern utilitarianism. The debate between so-called "virtue ethicists" like Mencius and defenders of rationalist approaches to ethics, such as utilitarians or deontologists, has raged unabated since the fourth century b.c.e., so the fact that contemporary science can weigh in—perhaps decisively—on one side of the

debate represents an enormous step forward. For more on this topic, see Slingerland (2011a).

117   *The existence of discrete, innate moral emotions.* For an interesting discussion of parallels between Mencian thought and evolutionary psychology, see Munro (2005).

117   *Most of this work has focused on empathy.* Preston and de Waal (2002).

117   *"Mirror neuron" system.* Rizzolatti, Fogassi, and Gallese (2001); Umiltà et al. (2001).

117   *Human psychopathy involves deficits in mirror-neuron and emotional regions of the brain.* Blair (2001).

118   *The usual rejection threshold hovering around 20 to 30 percent.* The findings of Henrich et al. (2006) from a study of subjects from five continents, representing the full range of political and economic styles, suggest the presence of a widespread tendency to punish unfair offers that is then calibrated culturally, with Ultimatum Game offer-rejection thresholds varying significantly from culture to culture.

118   *Neuroimaging studies of subjects playing the Ultimatum Game.* de Quervain et al. (2004).

118   *Behavior in this game may have a genetic basis.* Wallace et al. (2007).

118   *One study found that capuchin monkeys refuse food . . .* Brosnan and de Waal (2003).

118   *Conceptions of purity derived from disgust reactions.* On disgust as a moral emotion, see Rozin, Haidt, and Fincher (2009); on Mencian sprouts and evolved human psychology, see Flanagan and Williams (2010).

119   *The "neo-Humeans."* See, for example, Nichols (2004); Prinz (2007).

119   *The "rational tail" wagged by the "emotional dog."* From Haidt (2001); for some representative work by Haidt and his colleagues, see Haidt, Koller, and Dias (1993); Wheatley and Haidt (2005).

119   *Patients suffering from damage to the ventromedial prefrontal cortex (VMPFC).* Damasio (1994, 1999).

120   *A far cry from what we'd consider competent moral agents.* Some researchers have compared the situation of patients with damaged VMPFCs to that of alcoholics or compulsive gamblers, in that mere rational knowledge that something is harmful is not necessarily adequate to motivate a person to avoid getting into trouble. The problem with pathologically impulsive behavior may therefore be not too much emotion but rather not *enough* emotion.

121   *We find Mencius in the small state of Teng.* Mencius 3.A.4; Van Norden (2008: 68–73).

122   *A famous exchange between Mencius and a ruler he is trying to reform.* *Mencius* 1.A.7; Van Norden (2008: 7–10).

127   *The eight-track camp has the upper hand.* At the eight-track end, see Barsalou (1999) and Kosslyn, Thompson, and Ganis (2006); at the digital end, see Pylyshyn (2003); also see the responses to Barsalou (1999) to get a sense of the outlines of the debate.

127   *We think of our lives as journeys. . . .* On the pervasive role of metaphor in our everyday lives, see Lakoff and Johnson (1980, 1999) and Johnson (1987). For a good review of the broader research on thought as imagistic, see Kosslyn, Thompson, and Ganis (2006).

128   *Lead him through some guided exercises.* For more on Mencian imaginative extension, see Ivanhoe (2002).

128   *"The class ends, the rider gets back on the elephant . . ."* Haidt (2005: 165).

129   *The general standing in for the mind (cold cognition) and the troops for the body (hot cognition).* *Mencius* 2.A.2; Van Norden (2008: 38).

129   *"I say the eyes by themselves cannot perceive it. . . ."* *Xunzi*, ch. 20; Knoblock (1994: 85).

130   *"When such joy is born, it cannot be stopped. . . ."* *Mencius* 4.A.27; Van Norden (2008: 101).

130   *Bodily strength should be expended solely on practical tasks, like agriculture or manufacturing.* In fact, there is an entire chapter in the *Mozi*, the record of the teachings of the founder of the Mohist school, entitled "Against Music."

131   *In a series of exchanges, Gaozi proposes metaphors . . .* *Mencius* 6.A.1–2; Van Norden (2008: 143–44).

132   *The futility of Gaozi's carving-and-polishing strategy.* According to ancient library catalogues there once existed a book called the *Gaozi*, but it was lost a long time ago. It's a shame, because we never get to hear Gaozi's response to Mencius's redirection of his images.

132   *It would be better to not have the sacred Book of History at all. . . .* *Mencius* 7.B.3; Van Norden (2008: 185).

132   *A kind of "trellis" for one's moral sprouts.* Ivanhoe (1990: 94).

133   *In one passage, we find him similarly encouraged. . . .* *Mencius* 1.B.1; Van Norden (2008: 16–17).

135   *"What is the difficulty? . . ."* *Mencius* 6.B.2; Van Norden (2008: 159–60).

6. FORGET ABOUT IT

138   *The purported author of the text bearing his name.* Although the *Zhuangzi*
       is traditionally attributed to a figure of that name, there is very little his-
       torical evidence supporting this claim. The first seven chapters of the
       received text (the "Inner Chapters") do, however, appear to be written
       by a single (brilliant) person, whom we might as well call "Zhuangzi."
       The rest of the book is a very mixed bag, probably containing material
       from different periods and even different schools of thought. In dis-
       cussing the *Zhuangzi*, I'll be focusing on the Inner Chapters and other
       chapters felt by scholars to be from the same general school (these are
       sometimes referred to as the "School of Zhuangzi" chapters).

139   *There is clearly genuine affection between these two men.* Huizi's death is
       one of the few events that ever appears to shake Zhuangzi up; he com-
       plains that, with Huizi gone, he "has no one left to talk to" (*Zhuangzi*,
       ch. 24; B. Watson 1968: 269). Even the death of Zhuangzi's own wife
       didn't have such an impact (*Zhuangzi*, ch. 18; B. Watson 1968: 191–92).

139   *"When I planted them they grew into enormous gourds. . . ." Zhuangzi*, ch.
       1; B. Watson (1968: 34–35).

139   *"Categorical inflexibility."* The villain in these cases, as one might suspect
       by now, is the lateral PFC, seat of cold cognition. It's significant that
       young children are much less prone to this condition. They haven't yet
       had time to grow underbrush in their heads, and they are still able to
       view an object in all its potentiality, rather than slotted into a particu-
       lar category. On the lateral PFC as the locus of categorical inflexibility,
       see Thompson-Schill, Ramscar, and Chrysikou (2009); on children, see
       German and Defeyter (2000); on "functional fixedness" in a small-scale
       society, see German and Barrett (2005).

140   *"When people are asleep, their spirits wander off. . . ." Zhuangzi*, ch. 2; B.
       Watson (1968: 37).

141   *The challenges of finding happiness in civilized life have not changed much
       over the millennia.* In fact, we find a strikingly similar diagnosis of
       the sickness of the "modern" age—actually, early-twentieth-century
       Europe—in the writings of the great Jewish theologian Martin Buber
       (1878–1965). Buber's magnum opus, *Ich und Du* (1923)—typically trans-
       lated as *I and Thou*, but perhaps better as *I and You*—focuses on what
       he saw as the rising materialism of the modern world, which he thought
       brought with it a corresponding shallowness of interpersonal relation-
       ships and decline in genuine spirituality. With the rise of science and

worldwide capitalism, Buber saw a danger that our lives were becoming dominated by relentless striving and purely instrumentalist reasoning, our values diluted by objectification and cheap gratification. His description of the modern "capricious man" strongly echoes the "petty understanding" passage from the *Zhuangzi*: "He constantly interferes, and always with some specific goal, in order to 'let it happen.' ... The unbelieving marrow of the capricious man cannot take anything seriously other than faithlessness and haphazard seeking, setting up trivial goals and then scheming how to attain them. His world is devoid of true sacrifice and grace, genuine relationships or the ability to just live in the moment. He lives rather in a world that consists of nothing but ends and means. It could not be otherwise, and its name is doom" (Buber 1923/1985: 58–59, translation modified from Kaufmann [Buber 1970: 109–10]). Interestingly, Buber was familiar with the *Zhuangzi* and even wrote a little-known book about him, *Talks and Parables of Zhuangzi* (1910). Buber appears to have been much taken with *wu-wei*, and something very much like *wu-wei* seems to have informed his later ideal of the pure, genuine I-You relationship—contrasted by Buber with the instrumental, exploitative "I-It" relationship that he felt was becoming all too common. In fact, all sorts of influences from both Zhuangzi and Laozi glimmer throughout *Ich und Du*, though they are rarely noted because typically Buber scholars know very little, if anything, about the Chinese material or even Buber's connections to Chinese thought. For one exception, see Herman (1996).

142   *"If you're committed to something being 'right,' ..."* *Zhuangzi*, ch. 2; B. Watson (1968: 39–40).

143   *With Confucius actually espousing very Zhuangzian (and quite un-Confucian) ideas.* There is some debate about the significance of Zhuangzi using Confucius as his mouthpiece, some feeling it's just a way of poking fun at Confucius, others that it reflects a degree of sympathy for Confucius's position. This story appears in *Zhuangzi*, ch. 4; B. Watson (1968: 56–58).

146   *Something that is best achieved when we can weaken the hold of the conscious mind.* Cognitive control is, of course, crucial when it comes to maintaining focus in goal-directed activities or overriding undesirable, automatic tendencies in the pursuit of other, more abstract priorities. Recall the Stroop task from chapter 1, where our conscious minds had been instructed to name the case the word was printed in, not the word itself. When presented with *upper*, successful performance required

our cognitive control regions to override our automatic tendency to read it as "upper," and to direct our speech centers to produce "lower" instead. This is an extremely useful ability. It involves a trade-off, however. When networks involving regions like the lateral PFC are controlling our behavior, certain types of creativity—particularly those that require cognitive flexibility—seem to suffer.

146 *Adults with PFC damage tend to do better on such tasks than healthy controls.* See the literature review in Thompson-Schill, Ramscar, and Chrysikou (2009).

146 *One recent study asked subjects to perform a Remote Associates Test.* Jarosz, Colflesh, and Wiley (2012).

147 *Brief distractions appear to enhance both problem-solving ability and skilled performance on physical tasks.* On creativity and unconscious thought, see Zhong, Dijksterhuis, and Galinsky (2008); also see the work of Jonathan Schooler and colleagues on "mind wandering" and creativity (Smallwood et al. 2008); on distraction and physical tasks, see Jefferies et al. (2008).

147 *Another imaginary dialogue between Yan Hui and Confucius. Zhuangzi,* ch. 19; B. Watson (1968: 200–201).

148 *Psychological literature on the phenomenon of "choking."* See Beilock et al. (2002) and DeCaro et al. (2011) for helpful academic reviews and Beilock (2010) for an accessible book-length survey.

149 *"Paradoxical performance."* Lewis and Linder (1997).

149 *In one classic experiment . . .* Baumeister (1984).

149 *A more recent study of expert baseball players.* Gray, Wiebusch, and Akol (2004).

149 *Another Confucius–Han Yui dialogue . . . that appears a couple chapters later. Zhuangzi,* ch. 6; B. Watson (1968: 90–91).

150 *"Harmonize myself with the Way. . . ."* Literally "the Great Thoroughfare," here serving as a synonym for "the Way."

150 *"Vacant and dazed, as if he had lost his companion." Zhuangzi,* ch. 2; B. Watson (1968: 36–37).

152 *"There is a point at which something begins. . . ." Zhuangzi,* ch. 2; B. Watson (1968: 43).

153 *Meant as spiritual* therapy, *not religious doctrine.* On Zhuangzi's "therapeutic skepticism," see Ivanhoe (1996: 200–201).

153 *One famous* koan *story.* "Chaozhou (Joshu) Washes a Bowl," from the collection *Gateless Gate,* compiled by the Chinese monk Wumen Huikai (1183–1260).

154   *If you are out boating on a lake.* Zhuangzi, ch. 20; B. Watson (1968: 212).

155   *One story tells of a man who is training fighting roosters for a king.* Zhuangzi, ch. 19; B. Watson (1968: 204).

156   *A story about a monkey trainer.* Zhuangzi, ch. 2; B. Watson (1968: 41).

156   *"Collect taxes from morning to night without meeting the slightest resistance."* Zhuangzi, ch. 20; Watson (1968: 212–13). There is an interesting and revealing parallel here in the writings of the Danish existentialist Søren Kierkegaard. Commenting on how anyone you meet in the street might be a "knight of faith," he exclaims, "Good lord, is this the man? Is it really he? Why, he looks like a tax collector!" (Kierkegaard 1954: 49).

157   *"If its eyes do not spot a suitable place, it will not look twice. . . ."* Zhuangzi, ch. 20; B. Watson (1968: 218).

157   *"The True Person of ancient times slept without dreaming. . . ."* Zhuangzi, ch. 6; B. Watson (1968: 77–78).

157   *"Now, as for what most people do. . . ."* Zhuangzi, ch. 18; B. Watson (1968: 191).

158   *"Once the pivot is centered in its socket . . ."* Zhuangzi, ch. 2; B. Watson (1968: 40).

158   *"Do not serve as an embodier of fame . . ."* Zhuangzi, ch. 7; B. Watson (1968: 97).

159   *"Human on the outside, Heavenly on the inside."* Zhuangzi, ch. 17; B. Watson (1968: 182).

159   *"Ugly enough to astound the world."* Zhuangzi, ch. 5; B. Watson (1968: 72–73).

160   *One story in the Zhuangzi concerns a man who has suffered foot amputation.* Zhuangzi, ch. 5; B. Watson (1968: 70–71).

160   *The "social contagion" effect documented by psychologists.* For a recent review, see Christakis and Fowler (2012).

161   *Fish "forgetting about each other's presence . . ."* Zhuangzi, ch. 6; B. Watson (1968: 80).

161   *In one of their many dialogues.* Zhuangzi, ch. 5; B. Watson (1968: 75–76).

163   *"Use to the fullest what you have received from Heaven."* Zhuangzi, ch. 7; B. Watson (1968: 97).

163   *"Piss and shit."* A quite literal translation, and deliberately intended by Zhuangzi to shock. Zhuangzi, ch. 21; B. Watson (1968: 241).

164   *"Zhuangzi was obsessed by the 'Heavenly' . . ."* Xunzi, ch. 21; Knoblock (1994: 102).

164   *"Words are not just wind. . . ."* Zhuangzi, ch. 2; B. Watson (1968: 39).

165   *One imaginary dialogue between Confucius and Laozi. Zhuangzi,* ch. 21;
        B. Watson (1968: 226).

7. THE PARADOX OF *WU-WEI*

168   *This is the paradox of* wu-wei. I coined this term in my 1998 PhD dis-
        sertation, inspired by an essay by David Nivison on "the paradox of
        virtue" (Nivison 1996: 31–43), and I elaborate on the paradox in Sling-
        erland (2000a, 2003b), with updated evidence in Slingerland (2008).
        Since then, an increasing number of scholars have also been drawn to
        tensions in early Chinese thought that look very much like the paradox
        of virtue or paradox of *wu-wei,* exploring this theme with varying de-
        grees of awareness of prior scholarship on the subject (see, e.g., Meyer
        2008; Graziani 2009; Bruya 2010; De Prycker 2011).

169   *We all know that trying is bad.* For more on the sudden-gradual debate,
        see Gregory (1987).

170   *"Strictly speaking, any effort we make is not good for our practice. . . ."*
        Suzuki (1970: 37, 100).

170   *"Christian version of the paradox of Plato's* Meno *. . ."* MacIntyre (1990);
        see Slingerland (2003b) for more on the Meno problem and Aristotle.

171   *The Hindu* Bhagavad Gita *centers on the mystery of* karma yoga. Embree
        (1988: 281–86).

171   *"Super-good and pure of heart . . ."* Barrows and Blackall (2008: 29).

173   *Paradox of virtue.* Nivison (1996: 17–56). Nivison was the PhD adviser
        of my own PhD adviser, P. J. Ivanhoe, and so my intellectual grand-
        father, and I've been influenced by both scholars' work at a variety of
        levels.

173   *"The king's offer of self-sacrifice, ideally, has this result. . . ."* Nivison (1996: 23).

174   *For most of the history of our species.* Klein (1989). More recent work on
        human sociality (Hill et al. 2011; Chudek, Zhao, and Henrich, forth-
        coming) has suggested that the cultural shaping of human prosocial
        psychology may predate our transition to city life: since hunter-gatherer
        band composition can include a high proportion of unrelated individu-
        als, these bands often interact in much larger tribal networks, and one-
        off interactions with strangers are by no means rare in nonagricultural
        peoples. In this case, the challenges involved in making the transition
        to civilized, urban living represents merely one particularly dramatic
        step in the cultural-genetic evolution of our uniquely social species.

174 *A variety of innate psychological adaptations tailor-made for this kind of small-group lifestyle.* For an introduction to these sorts of arguments, see the essays collected in Barkow, Cosmides, and Tooby (1992) and Buss (2005).

176 *"The strategic role of the emotions."* Frank (1988, 2001).

177 *A host of cooperation scenarios . . . that simply cannot be navigated by rational calculation alone.* Frank (1988: 255).

178 *The conscious pursuit of self-interest is actually incompatible with its attainment.* Frank (1988: ix); also see the essays gathered in a seminal volume edited by Randolph Nesse (2001) and Joyce, Sterelny, and Calcott (2013).

178 *They're deeply concerned with the paradox of* wu-wei. Slingerland (2008).

179 *"A father's being treated with filiality . . . is not a matter of exerting effort."* Yucong 3.8.

179 *"If you try to be filial, this is not true filiality. . . ."* Yucong 1.55–58.

180 *The commitment strategy scenario will inevitably result in a stable mix of "cooperators" . . . and "defectors."* For a helpful, accessible, and short overview of the evolutionary literature on cooperation, see Pennisi (2009).

181 *Peacocks, for instance, have ridiculously large, colorful tails.* Zahavi and Zahavi (1997).

181 *"I've got a big engine, don't try to race me."* There's some evidence that stotting works. A lion presented with impressive stotting goes off in search of easier prey, saving both the lion and gazelle a lot of trouble. Hormone levels and irrepressible competitiveness mean that engine-size signaling in human male teenagers actually functions more as an invitation than a warning, but probably because drag racing merely involves pushing a pedal to the metal and has minimal survival consequences. If teenagers had to race each other on *foot*, and the loser faced starvation or being eaten alive, you can bet you'd see a lot more backing down on the starting line.

181 *Unfakeable signals are also used by people to broadcast social reliability.* See Maynard Smith and Harper (2003) on signaling and cooperation in a variety of species.

182 *It's a sign of intense and sincere commitment . . . precisely because it's permanent.* When these values are picked up by isolated groups of suburban teenagers from death metal rock lyrics, the results may appear somewhat capricious and foolish. (I suspect that my colleague's son, now an accomplished professional in his thirties, regrets at least one or two of those tattoos.) It's important to see, though, that this pow-

erful desire to dramatically signal one's membership in a particular group—and thereby one's embrace of *its* values and rejection of *others'* values—is a common human impulse. By permanently marking yourself, you demonstrate your commitment to your team and also join your fate to mine in an irrevocable way. There's some evidence that the prevalence of *permanent* marking practices (as opposed to temporary marking, like wearing a certain type of clothing) correlates with the intensity of intergroup warfare. In areas where tribes are constantly warring against neighboring tribes, the pressure to cut down on potential defection means that the practice of branding people with permanent group markers is more likely to arise and be passed on (Sosis, Kress, and Boster 2007).

182   *Any foreign language learned as an adult will be spoken with a noticeable accent.* There are a variety of estimates (ranging from age six to sixteen) of when the "critical period" for accent-free language acquisition ends, which in any case seems to vary by individual. It might be better to view it as a merely "sensitive" period, since the rise in noticeable accent as a function of age of first exposure seems to be linear rather than having a sharp jump at any particular age. See Piske, MacKay, and Flege (2001) for a review.

182   *If you didn't grow up with us, you can't speak like us.* Work by Katherine Kinzler and colleagues demonstrates that accent is a very salient group marker for children: children selectively trust and cooperate with people who share their native accent, an effect that—interestingly—trumps physical appearance clues such as race (Kinzler, Dupoux, and Spelke 2007; Kinzler, Corriveau, and Harris 2011).

182   *All groups use their own versions of the shibboleth.* Americans during World War II famously used knowledge of baseball as a shibboleth for separating real Americans from foreign agents posing as Americans. Maybe you speak perfect American English because you're a German raised in a bilingual environment, but who won the 1939 World Series? This particular shibboleth is unfakeable in a way that's not dissimilar to accent: if you haven't grown up in my cultural environment, there's simply no way you could share my body of baseball knowledge. Living in Rome on sabbatical, I confronted a less high-stakes, but structurally similar, problem in my ability to mingle with local Romans and improve my Italian: if you don't know the ins-and-outs of Serie A football (soccer), you're probably not worth wasting time talking to. On the other hand, if you *can* speak with some knowledge of the latest travails

of Lazio versus Roma, then maybe you're not so bad, despite the funny accent. Let me buy you a beer.

183   *One's willingness to put in the time and effort sends a strong signal of commitment to the group.* Norenzayan et al. (under review).

184   *Our bodies tend to tell the truth.* Bulbulia (2008); Schloss (2008).

184   *I've got a more or less 100 percent effective lie detector at my disposal.* Interestingly, Charles Darwin, the great naturalist and formulizer of evolutionary theory, noted the same phenomenon in one of his own children. At one point he talks about a subtle facial expression he discerned in one of his children who was clearly lying, characterizing it "as unnatural brightness of the eyes . . . an odd, affected manner, impossible to describe" (Darwin 1872/1998: 261). Darwin's interest in his child's "odd, affected manner" was part of a greater interest in emotional expressions across animal species; his book is very much worth reading today for its profound insight into the similarity in emotional expressions and their possible evolutionary role in signaling states of mind.

184   *The psychologist Paul Ekman.* Ekman (1985).

186   *We can determine, almost instantly, others' personality traits. . . .* There is some evidence that one of the things that can be read in this way is the presence of hormones with desirable or undesirable social consequences. For instance, recent work has suggested that testosterone level, and therefore propensity toward violence, can be quickly and accurately determined from a thin slice of facial expressions and bodily clues (Stillman, Maner, and Baumeister 2010). At the opposite end of the social desirability spectrum, other recent studies have suggested that subjects can pick out people with a genotype variation that makes them more likely to have high levels of oxytocin, a key hormone when it comes to human trust, cooperation, and empathy (Kosfeld et al. 2005; Kogan et al. 2011).

186   *One study that has proven enormously useful to law enforcement agencies.* Vrij et al. (2008).

188   *The pupils of your eyes become slightly dilated.* Hess (1965), cited and discussed in Kahneman (2011: 32–38).

188   *A recent brain imaging study.* Greene and Paxton (2009).

189   *People are spontaneously generous if forced to make instant decisions.* Rand, Greene, and Nowak (2012); cf. a recent review in Tomasello (2012) and work by Valdesolo and DeSteno (2008) that suggests that moral hypocrisy is a conscious process that can be short-circuited by cognitive load.

190   *"How can a person's true character hide?"* Analects 2.10; Slingerland (2003a: 11).

190   *"The True Man of ancient times slept without dreaming. . . ." Zhuangzi*, ch. 6; B. Watson (1968: 77–79). This is an exceedingly difficult passage, and my translation is relying heavily on Burton Watson, who in turn is following the Japanese commentator Fukunaga Mitsuji.

191   *"The traditional saying 'Make your thoughts sincere' . . ."* For a full English translation of this text, see Chan (1963: 84–94).

192   *When we're dancing, singing, drinking, and playing.* This is why thinkers around the world and throughout history have seen the connection between spontaneity, commitment, and trust as central to human communal life. One very revealing example is the concept of "play" (*Spiel*) developed by the German thinker Hans-Georg Gadamer (1900–2002). Gadamerian play shares all of the salient features of *wu-wei*: effortlessness, loss of intentionality, and a feeling of being absorbed into something larger than oneself. For Gadamer, play involves a kind of "ecstatic self-forgetfulness" where one is completely absorbed and carried away by the structure of the game, allowing a transcendence of the limited, selfish ego (Gadamer 2004). The connection between play, religion, and commitment to the group also appears in the work of the Dutch scholar Johan Huizinga (1872–1945), whose *Homo Ludens: A Study of the Play Element of Culture* (Huizinga 1939/1955) argues that the instinct to play is central to our ability to come together into cultures. Like Gadamer, he sees the key element of play to be this surrender of the self to something bigger than the self. "In the form and function of play," Huizinga remarks, "man's consciousness that he is embedded in a sacred order of things finds its first, highest, and holiest expression" (1939/1955: 17). For a group of individuals to unite into something more than just a collective of independent units, they need to *play* together, to lose themselves in a new, shared reality.

193   *Stable dispositions to perform socially desirable actions.* For an introduction to virtue ethics, and how it differs from the cold-cognition styles of ethics more dominant in recent Western thought, see, for example, MacIntyre (1985), or the essays collected in Crisp and Slote (1997). Van Norden (2007) and Angle (2009) have nice discussions that also connect these themes to the Chinese context.

194   *"I had two pairs of socks and I was still cold."* Goodman (2012).

194   *Evidence of sincerity and spontaneity in the moral realm inspires and moves us.* Subsequent coverage casting doubts upon the actual neediness and even moral character of the homeless man (Santora and Vadukul 2012; Jones 2012) merely serves as further evidence of what

might be called our "sincerity anxiety" when it comes to what appear to be moral acts.

194   *A distinction between physical skills (what he called "crafts") and virtues.* *Nicomachean Ethics* 1105a27–31; Aristotle (1999: 39–40). In fact, it was the crucial importance of internal motivation when it comes to virtue that led Aristotle to the view we noted in the beginning of this chapter, that "if we do what is just or temperate, we must already be just or temperate"—that is, it's impossible to train someone to acquire a virtue he or she did not already possess, at least in some incipient form.

196   *We'd demote it to a mere "problem" or "puzzle."* Colleagues of mine (including my former adviser!) have argued that this is precisely what we should do; see Fraser (2007) and Ivanhoe (2007b) for critiques of the idea that the paradox of *wu-wei* is a genuine paradox. See Slingerland (2008) for my academic response, as well as the helpful discussion in Knightly (2013). This book also constitutes a much more elaborate and scientifically grounded extension of the arguments made in Slingerland (2003b, 2008).

8. LEARNING FROM *WU-WEI*

199   *People are also born with liberal or conservative inclinations.* For a recent review article on genetics and the conservative-liberal divide, see Hatemi and McDermott (2012).

200   *Cold cognition begins to disrupt your performance.* See review and references in DeCaro and Beilock (2010).

200   *Creating a new, artificial nature as crucial for civilized life.* The British philosopher Alfred North Whitehead (1861–1947), for one, decried his culture's overemphasis on conscious reflection and self-control: "It is a profoundly erroneous truism, repeated by all copy-books and by eminent people making speeches, that we should cultivate the habit of thinking of what we are doing. The precise opposite is the case. Civilization advances by extending the number of operations which we can perform without thinking about them. Operations of thought are like cavalry charges in a battle—they are strictly limited in number, they require fresh horses, and must only be made at decisive moments" (Whitehead 1911, cited in Bargh and Chartrand 1999). The limitedness of cold cognition means that almost any behavior that we want to become fast and reliable needs to be automated in this fashion. The

French neuroscientist Alain Berthoz has coined the neologism *simplexité* ("simplex") to refer to the phenomenon whereby the brain compresses what was originally a multitude of effortful motor programs into one beautiful, smooth, "simplex" operation (Berthoz 2009). We simply cannot function in the world unless our hot cognition has been carved and polished in this way.

200   *The scholars of Confucianism Roger Ames and Henry Rosemont Jr.* . . . See, for instance, their introduction in Rosemont and Ames (2009); but also compare Ivanhoe (2008) for perspective on some of their stronger claims concerning the "otherness" of Confucian conceptions of the self.

201   *"Smart people who've thought about this . . ."* Brooks (2013b); for another Confucian-themed column, see also Brooks (2013a).

201   *"Ethical bootstrapping."* Sarkissian (2010); also see Hutton (2006).

201   *Seemingly trivial aspects of the social and physical environment can have profound effects on behavior.* For a popular introduction to the situationist literature, see Sommers (2012); see Slingerland (2011b) for a discussion of the situationist critique and Confucian ethics.

201   *A wake-up call to liberals such as myself.* Indeed, it's possible that cultural conservatives like William Bennett, who argue that many of today's social ills are caused by our rejection of tradition, may not be as crazy as we coastal big-city liberals are wont to think. There are some Confucian-style insights lurking in there among the flag-waving and chest-thumping. His *Book of Virtues* (Bennett 1993), for instance, looks very much like a traditional Confucian text, conveying moral virtues by means of traditional exemplars.

202   *Simply through an intensive course of training.* See Reber and Slingerland (2011) for a discussion and references.

202   *We can use the same techniques to foster our own particular set of values.* This may sound a bit like *fengshui*, but please don't call it that. (Possibly the single most annoying thing about being a professor of Chinese philosophy is being constantly asked about *fengshui*.) For those who have been somehow miraculously spared, the basic idea behind *fengshui* (literally, "wind and water") is that harmonizing the *qi* of your dwelling or business can bring prosperity and happiness. I don't put a whole lot of stock in *fengshui*, and because it arises after the period of Chinese philosophy that I study I'm not even professionally obligated to pretend to be interested in it. I would, though, argue that there is a kernel of truth lurking at the center of this otherwise wacky practice, and it actually has its roots in early Confucianism.

203   *Immersing yourself in reminders and environmental cues.* For more on the "power of habit," see Duhigg (2012).

203   *Letting the answer "pop out."* Smilek et al. (2006); M. Watson et al. (2010).

203   *Doing something else is often the best way to solve it.* Dijksterhuis and Meurs (2006); Zhong, Dijksterhuis, and Galinsky (2008).

203   *More effective than active suppression strategies.* Marcks and Woods (2005); Wegner (2011).

203   *"Governing a large state is like cooking a small fish."* Laozi, ch. 60; Ivanhoe (2003: 63).

204   *Sleep in, take a walk, go weed your garden.* On the "sleep on it" approach, see Ap Dijksterhuis's recent work on "Type 3" or "System 3" cognitive processes (Dijksterhuis et al. 2013).

205   *In a classic Modern Lovers track,* "Pablo Picasso," lyrics by Jonathan Richman, from the 1976 *The Modern Lovers* (Beserkley Records).

206   *As one recent article noted, however, there's very little evidence that these services work.* Schumpeter (2012).

206   *The attraction of Disneyland Shanghai. Economist* (2011).

207   *"Gangnam Style" being created by a committee of bureaucrats.* Indeed, Korean bureaucrats hoping to use "Gangnam Style" to promote Korea's image abroad seem to be scrambling to catch up with a craze they could never have predicted or orchestrated themselves (Fackler 2013).

207   *"Happiness is as a butterfly. . . ."* Schooler, Ariely, and Loewenstein (2003).

207   *A disruptive, negative effect on sensory pleasure.* Wilson and Schooler (1991); Wilson et al. (1993).

208   *Assets for both jam experts and wine professionals.* Melcher and Schooler (1996).

208   *The result is a refining feedback loop.* A popular urban myth (popular because it's elite-bashing, is my guess) holds that wine experts cannot tell the difference between even a white and a red in blind tastings, and that wine ranking and preferences all come down to social suggestion and marketing. Empirical studies suggest that novices are highly suggestible and variable in their ratings and descriptions but that experts are much less so, especially when it comes to styles of wine in which they have a great deal of experience. See Lehrer (2009) and Ashton (2012) for reviews.

208   *"When any . . . act of charity or of gratitude . . ."* Quoted in Haidt (2005: 195).

209 *Watching empathy-inspiring video clips increases generosity and caring be-havior.* The mediator seems to be oxytocin, a key hormone active in human trust and cooperation, with the causality going both ways. Shooting oxytocin up people's noses makes them more generous and trusting in social interactions (Kosfeld et al. 2005; Zak, Stanton, and Ahmadi 2007). Flipping this around, inducing empathy in people by showing them emotionally charged video clips increases their oxyto-cin levels, which in turn makes them more generous in subsequent economic games (Barraza and Zak 2009) and can even increase milk expression and caring behavior in nursing mothers (Silvers and Haidt 2008).

209 *Compared to novice meditators, the experts showed a markedly enhanced response.* . . . Lutz et al. (2008).

210 *A significant effect on children's ability to both experience and express compassion.* Ozawa-de Silva and Dodson-Lavelle (2011). Assessing actual measurable impact, however, awaits the results of a large-scale, properly randomized study that is currently under way; for more background and progress of this program, see Emory-Tibet Partnership (2013).

210 *Simple acts of generosity . . . can increase subjective feelings of happiness.* Sheldon and Lyubomirsky (2004); Dunn and Norton (2013).

210 *Keeping a gratitude journal . . . leads to increased compassion for others.* Emmons and McCullough (2003, 2004).

210 *Focusing conscious awareness on the mechanics of one's performance . . . has a disruptive effect on more experienced players.* Beilock et al. (2002).

211 *There are various hypotheses about why directing one's attention out, rather than in, is more effective.* See review in Wulf and Lewthwaite (2010). Also see the related work by Robin Jackson and colleagues, which sug-gests that athletes are more effective when they focus on strategic end goals or holistic bodily strategies rather than the actual details of their own performance, like which side of the foot to use while dribbling or how precisely to move their knees (Jackson, Ashford, and Norsworthy 2006).

211 *Focusing on the skill-relevant environment facilitates your ability.* . . . See Beilock (2010) for much more on the sports psychology literature on "choking," as well as some specific advice on how to "play 'outside your head' or at least outside your prefrontal cortex" (199).

211 *A growing empirical literature on this type of meditation.* See Shapiro, Schwartz, and Santerre (2005) and Siegel (2007) for reviews.

211 *Creating room for hot cognition to do its thing.* Austin (2001).

212   *Meditation has also been shown to increase self-esteem, empathy, trust.* Review in Shapiro, Schwartz, and Santerre (2005).

212   *Actually caring about the conversation ... is what's really important.* Mihaly Csikszentmihalyi provides some useful advice along these lines for someone who wants to join a party without appearing too withdrawn or too "slick." This is basically advice for how to socialize with *de*. He observes that a person in flow, "upon entering the room, would shift his attention away from himself to the party—the 'action system' he wishes to join" (Csikszentmihalyi 1990: 210–11). He would open himself up to observing the crowd, allowing himself to be drawn into conversations that seem interesting and to which he'd have something to contribute. When running into negative feedback, he would back off; when welcomed with positive feedback, he would advance. This sounds very much like Confucius or a Zhuangzian sage in action. Csikszentmihalyi's conclusion is that "only when a person's actions are appropriately matched with the opportunities of the action system does he truly become involved"—that is, it's all about a proper level of complexity and challenge. Our understanding of *wu-wei* would suggest, rather, that it's about losing the self and being absorbed into something larger.

212   *A shot of vodka might do the same.* Recalling the case of the baseball pitcher Steve Blass discussed in the Introduction, it's perhaps relevant here to note that one of the few times Blass was able to escape from his mental block and throw effectively in a real game was after having drunk a bottle of wine—a technique that he ultimately decided did not represent a helpful long-term solution to his problem (Hattenstone 2012). We've seen that alcohol and other intoxicants can function as a chemical short-cut to *wu-wei*, forcibly downregulating our conscious control regions. The artificial, temporary suppression of our conscious mind given to us by intoxicants may, in some situations, be precisely the little jump start that we need to push through the paradox of *wu-wei* in social situations. Interpersonal anxiety or awkwardness—an inability to enter *wu-wei* around other people—is typically overcome by shifting attention from yourself, allowing you to get caught up in the flow of the conversation or performance. It's the first step that's the hardest, although if you can somehow *begin* to relax, the relaxation will build upon itself in a positive loop. A bottle of wine or shot of vodka is probably not a useful option for someone who needs to pitch nine innings of baseball or return a one-hundred-miles-per-hour serve, but it might be just what the doctor ordered for an actress with stage fright, a stranger

insecure about joining a cocktail party conversation, or—as I myself found, at age twenty-six—a nervous graduate student about to give his first-ever lecture (on Zhuangzian *wu-wei*, no less!) to an auditorium full of people.

212　　*From abstract models of human cognition toward more embodied ones.* For classic works, see Neisser (1976) and Gibson (1979); for more recent, see Clark (1997) and Noë (2004).

213　　*Many trends can be picked out only by applying statistical tools.* Kahneman (2003, 2011).

213　　*"Metaphors we live by."* Lakoff and Johnson (1980).

213　　*"Imagine a person who can recite the several hundred Odes by heart...."* *Analects* 13.5; Slingerland (2003a: 141).

214　　*Why modern science arose in the West and not in China.* The question of "why China never developed science" was one running through the massive project on science and technology in China initiated by Joseph Needham, explored in the now almost thirty-volume series Science and Civilization in China, begun in 1954; for a comparative discussion of science and reason in ancient China and ancient Greece, see Lloyd and Sivin (2002).

Bibliography

Alcorta, Candace Storey, and Richard Sosis. 2007. Why ritual works: A rejection of the by-product hypothesis. *Behavioral and Brain Sciences* 29 (6): 613–14.

Ames, Roger. 2011. *Confucian role ethics.* Hong Kong: Chinese University of Hong Kong Press.

Angle, Stephen. 2009. Defining "virtue ethics" and exploring virtues in a comparative context. *Dao* 8 (3): 297–304.

Ansfield, Matthew, Daniel Wegner, and Robin Bowser. 1996. Ironic effects of sleep urgency. *Behavior Research and Therapy* 34 (7): 523–31.

Aristotle. 1999. *Nicomachean ethics.* Translated by Terence Irwin. Indianapolis: Hackett.

Ascher, L. Michael, and Ralph Turner. 1980. A comparison of two methods for the administration of paradoxical intention. *Behavior Research and Therapy* 18 (2): 121–26.

Ashton, Robert. 2012. Reliability and consensus of experienced wine judges: Expertise within and between? *Journal of Wine Economics* 7 (1): 70–87.

Austin, James. 2001. *Zen and the brain.* Cambridge, MA: MIT Press.

Baier, Annette. 1994. *Moral prejudices: Essays on ethics.* Cambridge, MA: Harvard University Press.

Banich, Marie. 2009. Executive function: The search for an integrated account. *Current Directions in Psychological Science* 18 (2): 89–94.

Bargh, John. 1997. Reply to the commentaries. In *The automaticity of everyday life: Advances in social cognition,* edited by R. Wyer, 231–46. Mahwah, NJ: Erlbaum.

Bargh, John, and Tanya Chartrand. 1999. The unbearable automaticity of being. *American Psychologist* 54: 577–609.

Bargh, John, Mark Chen, and Lara Burrows. 1996. Automaticity of social behavior: Direct effects of trait construct and stereotype activation on action. *Journal of Personality & Social Psychology* 71 (2): 230–44.

Bargh, John, Kay Schwader, Sarah Hailey, Rebecca Dyer, and Erica Boothby. 2012. Automaticity in social-cognitive processes. *Topics in Cognitive Science* 16 (12): 593–605.

Barkow, Jerome H., Leda Cosmides, and John Tooby, eds. 1992. *The adapted mind: Evolutionary psychology and the generation of culture.* New York: Oxford University Press.

Barraza, Jorge, and Paul Zak. 2009. Empathy toward strangers triggers oxytocin release and subsequent generosity. *Annals of the New York Academy of Sciences* 1167: 182–89.

Barrett, Nathaniel F. 2011. Wuwei and flow: Comparative reflections on spirituality, transcendence, and skill in the *Zhuangzi. Philosophy East and West* 61 (4): 679–706.

Barrows, Annie, and Sophie Blackall. 2008. *Ivy and Bean: Bound to be bad.* San Francisco: Chronicle Books.

Barsalou, Lawrence. 1999. Perceptual symbol systems. *Behavioral and Brain Sciences* 22: 577–609.

Baumeister, Roy. 1984. Choking under pressure: Self-consciousness and paradoxical effects of incentives on skillful performance. *Journal of Personality and Social Psychology* 46: 610–20.

Baumeister, Roy, Ellen Bratslavsky, Mark Muraven, and Dianne Tice. 1998. Ego depletion: Is the active self a limited resource? *Journal of Personality and Social Psychology* 74: 1252–65.

Baumeister, Roy F., E. J. Masicampo, and Kathleen D. Vohs. 2010. Do conscious thoughts cause behavior? *Annual Review of Psychology* 62 (1): 331–61.

Baxter, John. 1999. *George Lucas: A biography.* London: HarperCollins.

Beilock, Sian. 2010. *Choke: What the secrets of the brain reveal about getting it right when you have to.* New York: Free Press.

Beilock, Sian, Thomas Carr, Clare MacMahon, and Janet Starkes. 2002. When paying attention becomes counterproductive: Impact of divided versus skill-focused attention on novice and experienced performance of sensorimotor skills. *Journal of Experimental Psychology* 8 (1): 6–16.

Bellah, Robert, Richard Madsen, William Sullivan, Ann Swidler, and Steven Tipton. 1996. *Habits of the heart: Individualism and commitment in American life.* Berkeley: University of California Press.

Bennett, William, ed. 1993. *The book of virtues: A treasury of great moral stories.* New York: Simon & Schuster.

Berthoz, Alain. 2006. *Emotion and reason: The cognitive neuroscience of decision-making.* New York: Oxford University Press.

———. 2009. *La simplexité.* Paris: Odile Jacob.

Blair, James. 2001. Neurocognitive models of aggression, the antisocial personality disorders, and psychopathy. *Journal of Neurology, Neurosurgery, and Psychiatry* 71: 727–31.

Blass, Steve, and Erik Sherman. 2013. *A Pirate for life*. New York: Triumph Books.

Bloom, Paul. 2004. *Descartes' baby: How the science of child development explains what makes us human*. New York: Basic Books.

Boehm, Christopher. 1999. *Hierarchy in the forest: The evolution of egalitarian behavior*. New York: Macmillan.

Boyce, Christopher. J., G. D. Brown, and S. C. Moore. 2010. Money and happiness: Rank of income, not income, affects life satisfaction. *Psychological Science* 21 (4): 471–75.

Boyd, Robert, Peter Richerson, and Joseph Henrich. 2011. The cultural niche: Why social learning is essential for human adaptation. *Proceedings of the National Academy of Sciences* 108, suppl. 2: 10918–25.

Braver, Todd. 2012. The variable nature of cognitive control: A dual mechanisms framework. *Trends in Cognitive Science* 16 (2): 106–13.

Brooks, David. 2013a. The learning virtues. *New York Times*, February 28, 2013.

———. 2013b. Suffering fools gladly. *New York Times*, January 3, 2013.

Brosnan, Sarah, and F. B. M. de Waal. 2003. Monkeys reject unequal pay. *Nature* 425: 297–99.

Bruya, Brian. 2010. The rehabilitation of spontaneity: A new approach in philosophy of action. *Philosophy East and West* 60 (2): 207–50.

Buber, Martin. 1923/1985. *Ich und Du*. Stuttgart: Philipp Reclam.

———. 1970. *I and Thou*. Translated by W. Kaufmann. New York: Charles Scribner's Sons.

Bulbulia, Joseph. 2008. Free love: Religious solidarity on the cheap. In *The evolution of religion: Studies, theories and critiques*, edited by J. Bulbulia, R. Sosis, E. Harris, R. Genet, C. Genet, and K. Wyman, 153–60. Santa Margarita, CA: Collins Foundation Press.

Buonarroti, Michelangelo. 1821. *Rime e prose di michelagnolo buonarroti, pittore, scultore, architetto, e poeta fiorentino*. Milan: G. Silvestri.

Buss, David, ed. 2005. *Handbook of evolutionary psychology*. New York: Wiley.

Cahn, B. Rael, and John Polich. 2006. Meditation states and traits: EEG, ERP, and neuroimaging studies. *Psychological Bulletin* 132 (2): 180–211.

Caine, Michael. 1990. *Acting in film: An actor's take on moviemaking*. New York: Applause Theatre Book Publishing.

Chan, Wing-Tsit. 1963. *A source book in Chinese philosophy*. Princeton: Princeton University Press.

Chase, William, and Herbert Simon. 1973. Perception in chess. *Cognitive Psychology* 4: 55–61.

Christakis, Nicholas, and James Fowler. 2012. Social contagion theory: Examining dynamic social networks and human behavior. *Statistics in Medicine.*

Chudek, Maciej, Wanying Zhao, and Joseph Henrich. 2013. Culture-gene co-evolution, large-scale cooperation and the shaping of human social psychology. In *Signaling, commitment, and emotion,* edited by R. Joyce, K. Sterelny, and B. Calcott. Cambridge, MA: MIT Press.

Clark, Andy. 1997. *Being there: Putting brain, body and world together again.* Cambridge, MA: MIT Press.

Collins, Paul. 2009. How the world's greatest golfer lost his game. *New Scientist,* June 13, 44–45.

Cook, Scott. 2004. The debate over coercive rulership and the "human way" in light of recently excavated Warring States texts. *Harvard Journal of Asiatic Studies* 64 (2): 399–440.

Crisp, Roger, and Michael Slote, eds. 1997. *Virtue ethics.* New York: Oxford University Press.

Csikszentmihalyi, Mihaly. 1988a. The flow experience and its significance for human psychology. In *Optimal experience: Psychological studies of flow in consciousness,* edited by M. Csikszentmihalyi and I. Csikszentmihalyi, 15–35. New York: Cambridge University Press.

———. 1988b. The future of flow. In *Optimal experience: Psychological studies of flow in consciousness,* edited by M. Csikszentmihalyi and I. Csikszentmihalyi, 364–83. New York: Cambridge University Press.

———. 1990. *Flow: The psychology of optimal experience.* New York: Harper and Row.

Csikszentmihalyi, Mihalyi, and Isabella Csikszentmihalyi. 1988a. Flow as a way of life: Introduction to part III. In *Optimal experience: Psychological studies of flow in consciousness,* edited by M. Csikszentmihalyi and I. Csikszentmihalyi, 183–92. New York: Cambridge University Press.

———, eds. 1988b. *Optimal experience: Psychological studies of flow in consciousness.* New York: Cambridge University Press.

Damasio, Antonio. 1994. *Descartes' error: Emotion, reason, and the human brain.* New York: G. P. Putnam's Sons.

———. 1999. *The feeling of what happens: Body and emotion in the making of consciousness.* New York: Harcourt.

Darwin, Charles. 1872/1998. *The expression of emotions in man and animals.* With introduction, afterword, and commentaries by Paul Ekman. New York: Oxford University Press.

DeCaro, Marci, and Sian Beilock. 2010. The benefits and perils of attentional control. In *Effortless attention: A new perspective on the cognitive science of attention and action*, edited by B. Bruya, 51–73. Cambridge, MA: MIT Press.

DeCaro, Marci, Robin Thomas, Neil Albert, and Sian Beilock. 2011. Choking under pressure: Multiple routes to skill failure. *Journal of Experimental Psychology* 140 (3): 390–406.

Dennett, Daniel. 1991. *Consciousness explained*. Boston: Little, Brown.

———. 1995. *Darwin's dangerous idea: Evolution and the meanings of life*. New York: Simon & Schuster.

De Prycker, Valérie, 2011. Unself-conscious control: Broadening the notion of control through experiences of flow and wu-wei. *Zygon* 46 (1): 5–25.

de Quervain, Dominique J., Urs Fischbacher, Valerie Treyer, Melanie Schellhammer, Ulrich Schnyder, Alfred Buck, and Ernst Fehr. 2004. The neural basis of altruistic punishment. *Science* 305: 1254–58.

de Sousa, Ronald. 1987. *The rationality of emotion*. Cambridge: Cambridge University Press.

Diener, Edward, Richard Lucas, and Christie Scollon. 2006. Beyond the hedonic treadmill: Revising the adaptation theory of well-being. *American Psychologist* 61 (4): 305–14.

Dietrich, Arne. 2003. Functional neuroanatomy of altered states of consciousness: The transient hypofrontality hypothesis. *Consciousness and Cognition* 12: 231–56.

———. 2004. Neurocognitive mechanisms underlying the experience of flow. *Consciousness and Cognition* 13: 746–61.

Dijksterhuis, Ap, and John Bargh. 2001. The perception-behavior expressway: Automatic effects of social perception on social behavior. In *Advances in experimental social psychology*, edited by M. P. Zanna, 1–40. San Diego: Academic Press.

Dijksterhuis, Ap, and A. Van Knippenberg. 1998. The relation between perception and behavior, or how to win a game of Trivial Pursuit. *Journal of Personality and Social Psychology* 74: 865–77.

Dijksterhuis, Ap, Maarten Bos, Loran Nordgren, and Rick van Baaren. 2006. On making the right choice: The deliberation-without-attention effect. *Science* 311: 1005–7.

Dijksterhuis, A., and T. Meurs. 2006. Where creativity resides: The generative power of unconscious thought. *Consciousness and Cognition* 15: 135–46.

Dijksterhuis, Ap, Madelijn Strick, Maarten Bos, and Loran Nordgren. 2013. Proposing system 3. Paper presented at the annual meeting of the Society for Social and Personality Psychology, January 19, New Orleans, LA.

Doyen, Stéphane, Olivier Klein, Cora-Elise Pichon, and Axel Cleeremans. 2012. Behavioral priming: It's all in the mind, but whose mind? *PLoS ONE* 7 (1): e29081.

Duhigg, Charles. 2012. *The power of habit: Why we do what we do in life and business*. New York: Random House.

Dunn, Elizabeth, and Michael Norton. 2013. *Happy money: The science of smarter spending*. New York: Simon & Schuster.

Durkheim, Émile. 1915/1965. *The elementary forms of the religious life*. Translated by J. W. Swain. New York: George Allen and Unwin.

Easterlin, Richard. 1974. Does economic growth improve the human lot? Some empirical evidence. In *Nations and households in economic growth: Essays in honour of Moses Abramovitz*, edited by P. David and M. Reder, 89–125. New York: Academic Press.

Economist. 2011. Sun Tzu and the art of soft power. December 17, 71–74.

Ekman, Paul. 1985. *Telling lies*. New York: Norton.

Embree, Ainslie, ed. 1988. *Sources of Indian tradition*. Vol. 1. *From the beginning to 1800*. 2nd ed. New York: Columbia University Press.

Emmons, Robert, and Michael McCullough. 2003. Counting blessings versus burdens: Experimental studies of gratitude and subjective well-being in daily life. *Journal of Personality and Social Psychology* 84: 377–89.

———, eds. 2004. *The psychology of gratitude*. New York: Oxford University Press.

Emory-Tibet Partnership. 2013. "Compassion Meditation Study." May 1. www.tibet.emory.edu/cbct/research.html.

Enns, James, and Geniva Liu. 2009. Attentional limits and freedom in visually guided action. In *Progress in brain research*, edited by N. Srinivasan, 183–94. Amsterdam: Elsevier.

Fackler, Martin. 2013. Trendy spot urges tourists to ride in and spend, "Gangnam style." *New York Times*, January 1, A1.

Flanagan, Owen. 2011. *The bodhisattva's brain: Buddhism naturalized*. New York: Bradford Books.

Flanagan, Owen, and Robert Anthony Williams. 2010. What does the modularity of morals have to do with ethics? Four moral sprouts plus or minus a few. *Topics in Cognitive Science* 2 (3): 430–53.

Frank, Robert H. 1988. *Passions within reason: The strategic role of the emotions*. New York: W. W. Norton.

———. 2001. Cooperation through emotional commitment. In *Evolution and the capacity for commitment*, edited by R. M. Nesse, 57–76. New York: Russell Sage.

Frankl, Viktor. 1965. *The doctor and the soul.* 2nd ed. New York: Knopf.

Fraser, Chris. 2007. On wu-wei as a unifying metaphor. *Philosophy East and West* 57 (1): 97–106.

Frederick, Shane, and George Loewenstein. 1999. Hedonic adaptation. In *Well-being: The foundations of hedonic psychology,* edited by D. Kahneman, E. Diener, and N. Schwartz, 302–29. New York: Russell Sage.

Freud, Sigmund. 1930/1969. *Civilization and its discontents.* Translated by J. Strachey. New York: W. W. Norton.

Gadamer, Hans-Georg. 2004. *Truth and method.* Translated by J. Weinsheimer and D. G. Marshall. 2nd rev. ed. New York: Crossroad.

Gailliot, Matthew, Roy Baumeister, C. Nathan DeWall, Jon Maner, E. Ashby Plant, Dianne Tice, Lauren Brewer, and Brandon Schmeichel. 2007. Self-control relies on glucose as a limited energy source: Willpower is more than a metaphor. *Journal of Personality and Social Psychology* 92 (2): 325–36.

Gazzaniga, Michael. 1998. *The mind's past.* Berkeley: University of California Press.

Gazzaniga, Michael S., G. R. Mangun, and Richard B. Ivry. 1998. *Cognitive neuroscience: The biology of the mind.* New York: W. W. Norton.

Geertz, Clifford. 1973. *The interpretation of cultures: Selected essays.* New York: Basic Books.

German, Tim P., and H. Clark Barrett. 2005. Functional fixedness in a technologically sparse culture. *Psychological Science* 16 (1): 1–5.

German, Tim P., and Margaret Anne Defeyter. 2000. Immunity to functional fixedness in young children. *Psychonomic Bulletin and Review* 7: 707–12.

Gibbs, Raymond. 2006. *Embodiment and cognitive science.* Cambridge: Cambridge University Press.

Gibson, James. 1979. *The ecological approach to visual perception.* Boston: Houghton Mifflin.

Gigerenzer, Gerd. 2002. *Adaptive thinking: Rationality in the real world.* New York: Oxford University Press.

Gladwell, Malcolm. 2005. *Blink: The power of thinking without thinking.* New York: Little, Brown.

Goffman, Erving. 1959. *The presentation of self in everyday life.* New York: Anchor.

Goodale, Melvyn, and A. David Milner. 2004. *Sight unseen: An exploration of conscious and unconscious vision.* New York: Oxford University Press.

Goodman, J. David. 2012. Photo of officer giving boots to barefoot man warms hearts online. *New York Times,* November 28.

Grafton, Scott, Eliot Hazeltine, and Richard Ivry. 1995. Functional mapping of sequence learning in normal humans. *Journal of Cognitive Neuroscience* 7: 497–510.

Graham, A. C. 1983. Taoist spontaneity and the dichotomy of "is" and "ought." In *Experimental essays on the Chuang-tzu*, edited by V. Mair, 2–23. Honolulu: University of Hawai'i Press.

Grahn, J., J. Parkinson, and A. Owen. 2009. The role of the basal ganglia in learning and memory: Neuropsychological studies. *Behavioral Brain Research* 199 (1): 53–60.

Gray, S. J., B. Wiebusch, and H. A. Akol. 2004. Cross-sectional growth of pastoralist Karimojong and Turkana children. *American Journal of Physical Anthropology* 125 (2): 193–202.

Graziani, Romain. 2009. Optimal states and self-defeating plans: The problem of intentionality in early Chinese self-cultivation. *Philosophy East and West* 59 (4): 440–66.

Greene, Joshua, and Joseph Paxton. 2009. Patterns of neural activity associated with honest and dishonest moral decisions. *Proceedings of the National Academy of Science* 106 (30): 12506–11.

Gregory, Peter, ed. 1987. *Sudden and gradual approaches to enlightenment in Chinese thought*. Honolulu: University of Hawai'i Press.

Grouios, George. 1992. Mental practice: A review. *Journal of Sport Behavior* 15 (1): 42–59.

Haidt, Jonathan. 2001. The emotional dog and its rational tail: A social intuitionist approach to moral judgment. *Psychological Review* 108 (4): 814–34.

———. 2005. *The happiness hypothesis: Finding modern truth in ancient wisdom*. New York: Basic Books.

———. 2012. *The righteous mind: Why good people are divided by politics and religion*. New York: Pantheon Books.

Haidt, Jonathan, Silvia Koller, and Maria Dias. 1993. Affect, culture, and morality, or is it wrong to eat your dog? *Journal of Personality and Social Psychology* 65: 613–28.

Harbach, Chad. 2011. *The art of fielding*. New York: Little, Brown.

Harris, Sam. 2004. *The end of faith: Religion, terror and the future of reason*. New York: W. W. Norton.

———. 2010. *The moral landscape: How science can determine human values*. New York: Free Press.

Hatemi, Peter K., and Rose McDermott. 2012. The genetics of politics: Discovery, challenges, and progress. *Trends in Genetics* 28 (10): 525–33.

Hattenstone, Simon. 2012. Choke therapy: The sports stars who blew their big chance. *Guardian*, June 23, 14.

Henrich, Joseph, Steven J. Heine, and Ara Norenzayan. 2010. The weirdest people in the world? *Behavioral and Brain Sciences* 33 (2–3): 61–83.

Henrich, Joseph, and Natalie Henrich. 2010. The evolution of cultural adaptations: Fijian food taboos protect against dangerous marine toxins. *Proceedings of the Royal Society: Biological Sciences* 277 (1701): 3715–24.

Henrich, Joseph, and Richard McElreath. 2003. The evolution of cultural evolution. *Evolutionary Anthropology* 12 (3): 123–35.

Henrich, Joseph, Richard McElreath, Abigail Barr, Jean Ensminger, Clark Barrett, Alexander Bolyanatz, Juan Camilo Cardenas, Michael Gurven, Edwin Gwako, Natalie Henrich, Carolyn Lesorogol, Frank Marlowe, David Tracer, and John Ziker. 2006. Costly punishment across human societies. *Science* 312: 1767–70.

Herman, Jonathan. 1996. *I and Tao.* Albany: State University of New York Press.

Hess, Eckhard. 1965. Attitude and pupil size. *Scientific American,* April, 46–54.

Hill, Kim, Robert Walker, Miran Božičević, James Eder, Thomas Headland, Barry Hewlett, A. Magdalena Hurtado, Frank Marlowe, Polly Wiessner, and Brian Wood. 2011. Co-residence patterns in hunter-gatherer societies show unique human social structure. *Science* 331: 1286–89.

Hobbes, Thomas. 1651/1985. *Leviathan.* London: Penguin.

Holden, Stephen. 1987. Carly Simon triumphs over her own panic. *New York Times,* June 17.

Horton, Robin. 1993. *Patterns of thought in Africa and the West.* New York: Cambridge University Press.

Huizinga, Johan. 1939/1955. Homo ludens: *A study of the play element in culture.* Boston: Beacon Press.

Hume, David. 1739/1888. *The treatise concerning human nature.* Edited by L. A. Selby-Bigge. Oxford: Oxford University Press.

Hutton, Eric. 2006. Character, situationism, and early Confucian thought. *Philosophical Studies* 127 (1): 37–58.

Ihara, Craig, and Ryan Nichols. 2013. Review of Ames, Roger, *Confucian Role Ethics. Dao: A Journal of Comparative Philosophy* 11: 521–26.

Isaacson, Walter. 2011. *Steve Jobs.* New York: Simon & Schuster.

Ivanhoe, P. J. 1990. *Ethics in the Confucian tradition.* Atlanta: Scholar's Press.

———. 1996. Was Zhuangzi a relativist? In *Essays on skepticism, relativism, and ethics in the Zhuangzi,* edited by P. Kjellberg and P. J. Ivanhoe, 196–214. Albany: State University of New York Press.

———. 1999. The concept of *de* ("virtue") in the Laozi. In *Religious and philosophical aspects of the Laozi,* edited by M. Csikszentmihalyi and P. J. Ivanhoe, 239–57. Albany: SUNY Press.

———. 2002. Confucian self-cultivation and Mengzi's notion of extension. In *Essays on the moral philosophy of Mengzi*, edited by X. Liu and P. Ivanhoe, 221–41. Cambridge, MA: Hackett.

———. 2003. *The Daodejing of Laozi*. Indianapolis: Hackett.

———. 2007a. Heaven as a source for ethical warrant in early Confucianism. *Dao* 6 (3): 211–20.

———. 2007b. The paradox of wu-wei? *Journal of Chinese Philosophy* 34 (2): 277–87.

———. 2008. The shade of Confucius: Social roles, ethical theory, and the self. In *Polishing the Chinese mirror: Essays in honor of Henry Rosemont Jr.*, edited by M. Chandler and R. Littlejohn, 34–49. New York: Global Scholarly Publications.

Jackson, Robin, Kelly Ashford, and Glen Norsworthy. 2006. Attentional focus, dispositional reinvestment, and skilled motor performance under pressure. *Journal of Sports and Exercise Psychology* 28: 49–68.

Jarosz, Andrew, Gregory Colflesh, and Jennifer Wiley. 2012. Uncorking the muse: Alcohol intoxication facilitates creative problem solving. *Consciousness and Cognition* 21: 487–93.

Jefferies, Lisa, Daniel Smilek, Eric Eich, and James Enns. 2008. Emotional valence and arousal interact in attentional control. *Psychological Science* 19 (3): 290–95.

Johnson, Mark. 1987. *The body in the mind: The bodily basis of meaning, imagination, and reason*. Chicago: University of Chicago Press.

Jones, Jonathan. 2012. The homeless man and the NYPD cop's boots: How a warm tale turns cold. *Guardian*, December 4.

Joyce, Richard, Kim Sterelny, and Brett Calcott, eds. 2013. *Signaling, commitment, and emotion*. Cambridge, MA: MIT Press.

Juslin, Patrik, and Daniel Västfjäll. 2008. Emotional responses to music: The need to consider underlying mechanisms. *Behavioral and Brain Sciences* 31: 559–621.

Kahneman, Daniel. 2003. A perspective on judgment and choice: Mapping bounded rationality. *American Psychologist* 58 (9): 697–720.

———. 2011. *Thinking, fast and slow*. New York: Farrar, Straus, Giroux.

Katz, S. H. , M. L. Hediger, and L. A. Valleroy. 1974. Traditional maize processing techniques in the new world: Traditional alkali processing enhances the nutritional quality of maize. *Science*, May 17, 765–73.

Keeley, Lawrence H. 1996. *War before civilization*. New York: Oxford University Press.

Kelley, W. M., C. N. Macrae, C. L. Wyland, S. Caglar, S. Inati, and T. F. Heatherton. 2002. Finding the self? An event-related fMRI study. *Journal of Cognitive Neuroscience* 14 (5): 785–94.

Khan, Uzma, and Ravi Dhar. 2007. Licensing effect in consumer choice. *Journal of Marketing Research* 43: 259–66.

Kierkegaard, Søren. 1954. *Fear and trembling* and *The sickness unto death*. Translated by W. Lowrie. Princeton: Princeton University Press.

Kinzler, Katherine, Kathleen Corriveau, and Paul Harris. 2011. Children's selective trust in native-accented speakers. *Developmental Science* 14: 106–11.

Kinzler, Katherine, Emmanuel Dupoux, and Elizabeth Spelke. 2007. The native language of social cognition. *Proceedings of the National Academy of Sciences* 104: 12577–80.

Klein, Richard G. 1989. *The human career: Human biological and cultural origins*. Chicago: University of Chicago Press.

Knightly, Nickolas. 2013. The paradox of *wuwei?* Yes and no. *Asian Philosophy* 23 (2): 115–136.

Knoblock, John. 1988. *Xunzi: A translation and study of the complete works*. Vol. 1. Stanford: Stanford University Press.

———. 1990. *Xunzi: A translation and study of the complete works*. Vol. 2. Stanford: Stanford University Press.

———. 1994. *Xunzi: A translation and study of the complete works*. Vol. 3. Stanford: Stanford University Press.

Kogan, Aleksandr, Laura Saslow, Emily Impett, Christopher Oveis, Dacher Keltner, and Sarina Saturn. 2011. Thin-slicing study of the oxytocin receptor (OXTR) gene and the evaluation and expression of the prosocial disposition. *Proceedings of the National Academy of Sciences* 108 (48): 19189–92.

Konvalinka, Ivana, Dimitris Xygalatas, Joseph Bulbulia, Uffe Schjødt, Else-Marie Jegindø, Sebastian Wallot, Guy Van Orden, and Andreas Roepstorff. 2011. Synchronized arousal between performers and related spectators in a fire-walking ritual. *Proceedings of the National Academy of Sciences* 108 (20): 8514–19.

Kosfeld, Michael, Markus Heinrichs, Paul Zak, Urs Fischbacher, and Ernst Fehr. 2005. Oxytocin increases trust in humans. *Nature* 432 (2): 673–76.

Kosslyn, Stephen, William Thompson, and Giorgio Ganis. 2006. *The case for mental imagery*. New York: Oxford University Press.

Lakoff, George, and Mark Johnson. 1980. *Metaphors we live by*. Chicago: University of Chicago Press.

———. 1999. *Philosophy in the flesh: The embodied mind and its challenge to Western thought*. New York: Basic Books.

Le Blanc, Steven. 2004. *Constant battles: Why we fight.* New York: St. Martin's Griffin.

LeDoux, Joseph. 1996. *The emotional brain: The mysterious underpinnings of emotional life.* New York: Simon & Schuster.

Lehrer, Adrienne. 2009. *Wine and conversation.* 2nd ed. New York: Oxford University Press.

Lewis, Brian, and Darwyn Linder. 1997. Thinking about choking? Attentional processes and paradoxical performance. *Personality and Social Psychology Bulletin* 23 (9): 937–44.

Limb, Charles, and Allen Braun. 2008. Neural substrates of spontaneous musical performance: An fMRI study of jazz improvisation. *PLoS ONE* 3 (2): e1679.

Lloyd, Geoffrey, and Nathan Sivin. 2002. *The way and the word: Science and medicine in early China and Greece.* New Haven: Yale University Press.

Lutz, Antoine, Julie Brefcynski-Lewis, Tom Johnstone, and Richard Davidson. 2008. Regulation of the neural circuitry of emotion by compassion meditation: Effects of meditative expertise. *PLoS One* 3 (3): e1897.

MacIntyre, Alasdair. 1985. *After virtue.* 2nd ed. London: Duckworth.

———. 1990. *Three rival versions of moral inquiry: Encyclopedia, genealogy, and history.* Notre Dame: University of Notre Dame Press.

Macrae, C. Neil, and Lucy Johnston. 1998. Help, I need somebody: Automatic action and inaction. *Social Cognition* 16 (4): 400–17.

Marcks, B., and D. Woods. 2005. A comparison of thought suppression to an acceptance-based technique in the management of personal intrusive thoughts: A controlled evaluation. *Behavioral Research and Therapy* 43: 433–45.

Maynard Smith, John, and David Harper. 2003. *Animal signals.* Oxford Series in Ecology and Evolution. Oxford: Oxford University Press.

McCauley, Robert. 2011. *Why religion is natural and science is not.* New York: Oxford University Press.

McNeill, William. 1995. *Keeping together in time: Dance and drill in human history.* Cambridge, MA: Harvard University Press.

Melcher, Joseph, and Jonathan Schooler. 1996. The misremembrance of wines past: Verbal and perceptual expertise differentially mediate verbal overshadowing of taste memory. *Journal of Memory and Language* 35: 231–45.

Meyer, Dirk. 2008. Writing meaning: Strategies of meaning-construction in early Chinese philosophical discourse. *Monumenta Serica* 56: 55–95.

Miller, Earl, and Jonathan Cohen. 2001. An integrative theory of prefrontal cortex function. *Annual Review of Neuroscience* 24: 167–202.

Monin, Benoît, and Dale Miller. 2001. Moral credentials and the expression of prejudice. *Journal of Personality and Social Psychology* 81 (1): 33–43.

Moore, G. E. 1903. *Principia ethica.* Cambridge: Cambridge University Press.

Munro, Donald. 2005. *A Chinese ethics for the new century: The Ch'ien Mu lectures in history and culture, and other essays on science and Confucian ethics.* Hong Kong: Chinese University of Hong Kong Press.

Muraven, Mark, Dianne Tice, and Roy Baumeister. 1998. Self-control as a limited resource: Regulatory depletion patterns. *Journal of Personality and Social Psychology* 74: 774–89.

Naccache, Lionel, Stanislas Dehaene, Laurent Cohen, Marie-Odile Habert, Elodie Guichart-Gomez, Damien Galanaud, and Jean-Claude Willer. 2005. Effortless control: Executive attention and conscious feeling of mental effort are dissociable. *Neuropsychologia* 43: 1318–28.

Nagell, K., Raquel Olguin, and Mark Tomasello. 1993. Processes of social learning in the tool use of chimpanzees *(Pan troglodytes)* and human children *(Homo sapiens). Journal of Comparative Psychology* 107: 174–86.

Neisser, Ulrich. 1976. *Cognition and reality: Principles and implications of cognitive psychology.* San Francisco: W. H. Freeman.

Nesse, Randolph, ed. 2001. *Evolution and the capacity for commitment.* New York: Russell Sage.

Nichols, Shaun. 2004. *Sentimental rules: On the natural foundations of moral judgment.* New York: Oxford University Press.

Nivison, David. 1996. *The ways of Confucianism.* Edited by B. Van Norden. La Salle, IL: Open Court.

Noë, Alva. 2004. *Action in perception.* Cambridge, MA: MIT Press.

Norenzayan, Ara, Azim Shariff, William Gervais, Aiyana Willard, Edward Slingerland, and Joseph Henrich. Under review. The cultural evolution of prosocial religions. *Behavioral and Brain Sciences.*

Ozawa-de Silva, Brendan, and Brooke Dodson-Lavelle. 2011. An education of heart and mind: Practical and theoretical issues in teaching cognitive-based compassion training to children. *Practical Matters,* no. 4, 1–28.

Pashler, Harold, Noriko Coburn, and Christine Harris. 2012. Priming of social distance? Failure to replicate effects on social and food judgments. *PLoS One* 7 (8): e42510.

Pennisi, Elizabeth. 2009. On the origin of cooperation. *Science* 325 (5945): 1196–99.

Pessoa, Luiz. 2005. To what extent are emotional visual stimuli processed without attention and awareness? *Current Opinion in Neurobiology* 15: 188–96.

Pinker, Steven. 2011. *The better angels of our nature: Why violence has declined.* New York: Viking.

Piske, Thorsten, Ian MacKay, and James Flege. 2001. Factors affecting degree of foreign accent in an L2: A review. *Journal of Phonetics* 29: 191–215.

Polanyi, Michael. 1967. *The tacit dimension.* Garden City, NJ: Doubleday.

Poldrack, Russ. 2006. Can cognitive processes be inferred from neuroimaging data? *Trends in Cognitive Science* 10: 59–63.

Poldrack, Russell, Fred Sabb, Karin Foerde, Sabrina Tom, Robert Asarnow, Susan Bookheimer, and Barbara Knowlton. 2005. The neural correlates of motor skill automaticity. *Journal of Neuroscience* 25: 5356–64.

Preston, Stephanie, and Frans de Waal. 2002. Empathy: Its ultimate and proximate bases. *Behavioral and Brain Sciences* 25: 1–72.

Prinz, Jesse. 2007. *The emotional construction of morals.* New York: Oxford University Press.

Pylyshyn, Zenon. 2003. Mental imagery: In search of a theory. *Behavioral and Brain Sciences* 25: 157–237.

Rand, David, Joshua Greene, and Martin Nowak. 2012. Spontaneous giving and calculated greed. *Nature* 489 (7416): 427–30.

Rappaport, Roy A. 1999. *Ritual and religion in the making of humanity.* Cambridge: Cambridge University Press.

Reber, Rolf, and Edward Slingerland. 2011. Confucius meets cognition: New answers to old questions. *Religion, Brain and Behavior* 1 (2): 135–45.

Redfield, Robert. 1953. *The primitive world and its transformations.* Ithaca: Cornell University Press.

Richerson, Peter J., and Robert Boyd. 2005. *Not by genes alone: How culture transformed human evolution.* Chicago: University of Chicago Press.

Rizzolatti, Giacomo, Leonardo Fogassi, and Vittorio Gallese. 2001. Neurophysiological mechanisms underlying the understanding and imitation of action. *Nature Reviews Neuroscience* 2: 661–70.

Rosemont, Henry, Jr., and Roger Ames. 2009. *The Chinese classic of family reverence: A philosophical translation of the Xiaojing.* Honolulu: University of Hawai'i Press.

Rozin, Paul, Jonathan Haidt, and Katrina Fincher. 2009. Psychology: From oral to moral. *Science* 323 (5918): 1179–80.

Ryle, Gilbert. 1949. *The concept of mind.* London: Hutchinson.

Sachdeva, Sonya, Rumen Iliev, and Douglas Medin. 2009. The paradox of moral self-regulation. *Psychological Science* 20 (4): 523–28.

Safire, William, and Leonard Safire, eds. 1989. *Words of wisdom: More good advice.* New York: Simon & Schuster.

Santora, Marc, and Alex Vadukul. 2012. Homeless man is grateful for officer's gift of boots. But he again is barefoot. *New York Times*, December 2.

Sarkissian, Hagop. 2010. Minor tweaks, major payoffs: The problems and promise of situationism in moral philosophy. *Philosopher's Imprint* 10 (9): 1–15.

Sato, Ikuya. 1988. Bosozoku: Flow in Japanese motorcycle gangs. In *Optimal experience: Psychological studies of flow in consciousness*, edited by M. Csikszentmihalyi and I. Csikszentmihalyi, 92–117. New York: Cambridge University Press.

Sayre-McCord, Geoff. 2012. Metaethics. In *The Stanford Encyclopedia of Philosophy* (Spring 2012 ed.), edited by E. Zalta. http://stanford.library.usyd.edu.au/archives/spr2012/.

Schloss, Jeffrey. 2008. He who laughs best: Involuntary religious affect as a solution to recursive cooperative defection. In *The evolution of religion: Studies, theories and critiques*, edited by J. Bulbulia, R. Sosis, E. Harris, R. Genet, C. Genet, and K. Wyman, 197–207. Santa Margarita, CA: Collins Foundation Press.

Schmeichel, Brandon, and Roy Baumeister. 2010. Effortful attention control. In *Effortless attention: A new perspective in the cognitive science of attention and action*, edited by B. Bruya, 29–47. Cambridge, MA: MIT Press.

Schooler, Jonathan, Dan Ariely, and George Loewenstein. 2003. The pursuit and assessment of happiness can be self-defeating. In *Psychology and economics*, edited by J. Carrillo and I. Brocas, 41–70. Oxford: Oxford University Press.

Schooler, Jonathan, and Tonya Engstler-Schooler. 1990. Verbal overshadowing of visual memories: Some things are better left unsaid. *Cognitive Psychology* 22 (1): 36–71.

Schumpeter. 2012. What's in a name? *Economist*, April 21, 69.

Seok, Bongrae. 2012. *Embodied moral psychology and Confucian philosophy*. New York: Rowman and Littlefield.

Shapiro, Shauna, Gary Schwartz, and Craig Santerre. 2005. Meditation and positive psychology. In *Handbook of positive psychology*, edited by C. R. Snyder and S. Lopez, 632–45. Oxford: Oxford University Press.

Sheldon, K., and S. Lyubomirsky. 2004. Achieving sustainable new happiness: Prospects, practices, and prescriptions. In *Positive psychology in practice*, edited by P. Linley and S. Joseph, 127–45. Hoboken, NJ: Wiley.

Siegel, Daniel. 2007. *The mindful brain: Reflection and attunement in the cultivation of well-being*. New York: W. W. Norton.

Silvers, Jennifer, and Jonathan Haidt. 2008. Moral elevation can induce lactation. *Emotion* 8 (2): 291–95.

Simon spanks away her stage fright. 2006. Contactmusic.com, August 29. www.contactmusic.com/news/simon-spanks-away-her-stage-fright_1006576.

Singer, Peter. 2011. *Practical ethics.* 3rd ed. New York: Cambridge University Press.

Slingerland, Edward. 1998. Effortless action: Wu-wei as spiritual ideal in early China. PhD diss., Stanford University.

———. 2000a. Effortless action: The Chinese spiritual ideal of wu-wei. *Journal of the American Academy of Religion* 68 (2): 293.

———. 2000b. Why philosophy is not "extra" in understanding the *Analects.* Review of *The Original Analects,* by E. Bruce Brooks and A. Taeko Brooks. *Philosophy East and West* 50 (1): 137–41.

———. 2003a. *Confucius: Analects: With selections from traditional commentaries.* Indianapolis: Hackett.

———. 2003b. *Effortless action: Wu-wei as conceptual metaphor and spiritual ideal in early China.* New York: Oxford University Press.

———. 2007. Confucius as secular savior? A problem with Enlightenments 0.5–2.0. Paper presented at the conference "Beyond Belief: Enlightenment 2.0," Salk Institute for Biological Studies, San Diego, CA, October 31–November 2, http://thesciencenetwork.org/programs/beyond-belief-enlightenment-2-0/edward-slingerland.

———. 2008. The problem of moral spontaneity in the Guodian corpus. *Dao: A Journal of Comparative Philosophy* 7 (3): 237–56.

———. 2011a. "Of what use are the *Odes?*" Cognitive science, virtue ethics, and early Confucian ethics. *Philosophy East and West* 61 (1): 80–109.

———. 2011b. The situationist critique and early Confucian virtue ethics. *Ethics* 121 (2): 390–419.

———. 2013. Body and mind in early China: An integrated humanities-science approach. *Journal of the American Academy of Religion* 81 (1): 6–55.

Sloman, Steven. 1996. The empirical case for two systems of reasoning. *Psychological Bulletin* 119: 3–22.

Smallwood, J., M. McSpadden, B. Luus, and Jonathan Schooler. 2008. When attention matters: The curious incident of the wandering mind. *Memory and Cognition* 36: 1144–50.

Smilek, Daniel, James Enns, John Eastwood, and Philip Merikle. 2006. Relax! Cognitive strategy influences visual search. *Visual Cognition* 14: 543–64.

Smith, Michael, Linda McEvoy, and Alan Gevins. 1999. Neurophysiological indices of strategy development and skill acquisition. *Cognitive Brain Research* 7: 389–404.

Smith, Natalie. 2001. Are indigenous people conservationists? Preliminary results from the Machiguenga of the Peruvian Amazon. *Rationality and Society* 13: 429–61.

Sommers, Sam. 2012. *Situations matter: Understanding how context transforms your world*. New York: Riverhead.

Sosis, Richard, Howard Kress, and James Boster. 2007. Scars for war: Evaluating alternative signaling explanations for cross-cultural variance in ritual costs. *Evolution and Human Behavior* 28: 234–47.

Sperber, Dan. 1996. *Explaining culture: A naturalistic approach*. Oxford: Blackwell.

Sports Illustrated. 2005. How it feels . . . to be on fire. February 21. http://sportsillustrated.cnn.com/vault/article/magazine/MAG1106119/index.htm.

Stanovich, Keith, and Richard West. 2000. Individual difference in reasoning: Implications for the rationality debate? *Behavioural and Brain Sciences* 23: 645–726.

Stern, Daniel. 1977. *The first relationship: Infant and mother*. Cambridge, MA: Harvard University Press.

Stillman, Tyler F., Jon K. Maner, and Roy F. Baumeister. 2010. A thin slice of violence: Distinguishing violent from nonviolent sex offenders at a glance. *Evolution and Human Behavior* 31 (4): 298–303.

Stout, Dietrich. 2010. The evolution of cognitive control. *Topics in Cognitive Science* 2 (4): 614–30.

Strack, Fritz, Leonard Martin, and Sabine Stepper. 1988. Inhibiting and facilitating conditions of the human smile: A nonobtrusive test of the facial feedback hypothesis. *Journal of Personality and Social Psychology* 54: 768–77.

Suzuki, Shunryu. 1970. *Zen mind, beginner's mind: Informal talks on Zen meditation and practice*. New York: Weatherhill.

Syed, Matthew. 2010. *Bounce: Mozart, Federer, Picasso, Beckham, and the science of success*. New York: Harper.

Taves, Ann. 2009. *Religious experience reconsidered: A building-block approach to the study of religion and other special things*. Princeton: Princeton University Press.

Taylor, Charles. 1989. *Sources of the self: The makings of modern identity*. Cambridge, MA: Harvard University Press.

———. 2007. *A secular age*. Cambridge, MA: Harvard University Press.

Thompson, Evan. 2007. *Mind in life: Biology, phenomenology, and the sciences of mind*. Cambridge, MA: Harvard University Press.

Thompson-Schill, Sharon, Michael Ramscar, and Evangelia Chrysikou. 2009. Cognition without control: When a little frontal lobe goes a long way. *Current Directions in Psychological Science* 18 (5): 259–63.

Thoreau, Henry David. 1854/1949. *Walden: An annotated edition*. Edited with foreword and notes by Walter Harding. Boston: Houghton Mifflin.

Tomasello, Michael. 2012. Why be nice? Better not to think about it. *Topics in Cognitive Science* 16 (12): 580–81.

Umiltà, Alessandra, Evelyne Kohler, Vittorio Gallese, Leonardo Fogassi, Luciano Fadiga, Christian Keysers, and Giacomo Rizzolatti. 2001. I know what you are doing: A neurophysiological study. *Neuron* 31: 155–65.

Valdesolo, Piercarlo, and David DeSteno. 2008. The duality of virtue: Deconstructing the moral hypocrite. *Journal of Experimental Social Psychology* 44: 1334–38.

Van Norden, Bryan. 2007. *Virtue ethics and consequentialism in early Chinese philosophy*. New York: Cambridge University Press.

———. 2008. *Mengzi: With selections from traditional commentaries*. Cambridge, MA: Hackett.

Varela, Francisco, Evan Thompson, and Eleanor Rosch. 1991. *The embodied mind: Cognitive science and human experience*. Cambridge, MA: MIT Press.

Vrij, A., S. A. Mann, R. P. Fisher, S. Leal, R. Milne, and R. Bull. 2008. Increasing cognitive load to facilitate lie detection: The benefit of recalling an event in reverse order. *Law and Human Behavior* 32: 253–65.

Wallace, Björn, David Cesarini, Paul Lichtenstein, and Magnus Johannesson. 2007. Heritability of ultimatum game responder behavior. *Proceedings of the National Academy of Science* 104: 15631–34.

Watson, Burton. 1968. *The complete works of Chuang Tzu*. New York: Columbia University Press.

Watson, Marcus, Allison Brennan, Alan Kingstone, and James Enns. 2010. Looking versus seeing: Strategies alter eye movements during visual search. *Psychonomic Bulletin and Review* 17 (4): 543–49.

Wegner, Daniel. 2002. *The illusion of conscious will*. Cambridge, MA: MIT Press.

———. 2009. How to think, say, or do precisely the worst thing for any occasion. *Science* 325: 48–50.

———. 2011. Setting free the bears: Escape from thought suppression. *American Psychologist* 66 (8): 671–80.

Wegner, Daniel, Matthew Ansfield, and Daniel Pilloff. 1998. The putt and the pendulum: Ironic effects on mental control of action. *Psychological Science* 9 (3): 196–99.

Wegner, Daniel, and John Bargh. 1998. Control and automaticity in social life. In *Handbook of social psychology*, edited by D. Gilbert, S. Fiske, and G. Lindzey, 446–96. Boston: McGraw-Hill.

Wheatley, Thalia, and Jonathan Haidt. 2005. Hypnotic disgust makes moral judgments more severe. *Psychological Science* 16: 780–84.

Whitehead, Alfred North. 1911. *An introduction to mathematics*. New York: Holt.

Wilson, Timothy. 2002. *Strangers to ourselves: Discovering the adaptive unconscious.* Cambridge, MA: Harvard University Press.

Wilson, Timothy, Dolores Kraft, and Dana Dunn. 1989. The disruptive effects of explaining attitudes: The moderating effect of knowledge about the attitude object. *Journal of Experimental Social Psychology* 25: 379–400.

Wilson, Timothy, Douglas Lisle, Jonathan Schooler, Sara Hodges, Kristen Klaaren, and Suzanne LaFleur. 1993. Introspecting about reasons can reduce post-choice satisfaction. *Personality and Social Psychology Bulletin* 19 (3): 331–39.

Wilson, Timothy, and Jonathan Schooler. 1991. Thinking too much: Introspection can reduce the quality of preferences and decisions. *Journal of Personality and Social Psychology* 60 (2): 181–92.

Wiltermuth, Scott, and Chip Heath. 2009. Synchrony and cooperation. *Psychological Science* 20 (1): 1–5.

Wulf, Gabriele, and Rebecca Lewthwaite. 2010. Effortless motor learning? An external focus of attention enhances movement effectiveness and efficiency. In *Effortless attention: A new perspective in attention and action,* edited by B. Bruya, 75–102. Cambridge, MA: MIT Press.

Yglesias, Matthew. 2011. Steve Jobs and the economics of place. *ThinkProgress,* November 10. http://thinkprogress.org/yglesias/2011/11/10/365937/steve-jobs -and-the-economics-of-place/.

Yong, Ed. 2012. Nobel laureate challenges psychologists to clean up their act. *Nature* 490 (7418).

Zahavi, Amotz, and Avishag Zahavi. 1997. *The handicap principle: A missing piece of Darwin's puzzle.* New York: Oxford University Press.

Zajonc, Robert. 1980. Feeling and thinking: Preferences need no inferences. *American Psychologist* 35: 151–75.

Zak, Paul, Angela Stanton, and Sheila Ahmadi. 2007. Oxytocin increases generosity in humans. *PLoS One* 11: e1128.

Zhong, C. B., A. Dijksterhuis, and A. D. Galinsky. 2008. The merits of unconscious thought in creativity. *Psychological Science* 19 (9): 912–18.

Index

About the Author

EDWARD SLINGERLAND is Professor of Asian Studies and Canada Research Chair in Chinese Thought and Embodied Cognition at the University of British Columbia. Educated at Princeton, Stanford, and the University of California, Berkeley, he is an internationally renowned expert in Chinese thought, comparative religion, and cognitive science. In addition to more than twenty academic journal articles in a range of fields, he has written several scholarly books, including *What Science Offers the Humanities* and a translation of the *Analects* of Confucius. He lives in Vancouver with his wife and daughter.